Nanoscale Science and Engineering Forum 2022

Held at the 2022 AIChE Annual Meeting

Phoenix, Arizona, USA
13-18 November 2022

ISBN: 978-1-7138-7890-2

Printed from e-media with permission by:

Curran Associates, Inc.
57 Morehouse Lane
Red Hook, NY 12571

AIChE
120 Wall Street, FL 23
New York, NY 10005-4020

Phone: (800) 242-4363
Fax: (203) 775-5177

www.aiche.org

Additional copies of this publication are available from:

Curran Associates, Inc.
57 Morehouse Lane
Red Hook, NY 12571 USA
Phone: 845-758-0400
Fax: 845-758-2633
Email: curran@proceedings.com
Web: www.proceedings.com

TABLE OF CONTENTS

NANOSCALE SCIENCE AND ENGINEERING FORUM DIVISION PLENARY

AREA PLENARY: CARBON NANOMATERIALS (INVITED TALKS)

PLENARY SESSION FOR NANOMATERIALS FOR ENERGY APPLICATIONS (INVITED TALKS)

NANOSCALE SCIENCE AND ENGINEERING FORUM I (ALL PAPERS)

CARBON NANOMATERIALS II: DISPERSION, SURFACE STRUCTURE, AND BIOINTERACTIONS

NANOTECHNOLOGY APPROACHES TO DIAGNOSTICS, IMPLANTS, TEMPLATING AND ASSEMBLY

POSTER SESSION: NANOSCALE SCIENCE AND ENGINEERING FORUM

POSTER SESSION: NSEF GRADUATE STUDENT POSTER COMPETITION

NANOMATERIALS FOR ENERGY STORAGE

NANOMATERIALS FOR ENERGY CONVERSION

NANOMATERIALS FOR GENERAL ELECTROCHEMICAL PHENOMENA

CARBON NANOMATERIALS I: DISPERSION, SURFACE STRUCTURE, AND BIOINTERACTIONS

NANOSCALE SCIENCE AND ENGINEERING FORUM II (ALL PAPERS)

NANOPARTICLE DRUG DELIVERY SYSTEMS

457a Molecular Engineering of Biomimetic Condensates and Polyelectrolyte Complex Micelles

Lorraine Leon

Biological condensates (membraneless organelles) are liquid droplets consisting of proteins and nucleic acids that form within cells and carefully regulate biochemical reactions. These droplets are dynamic, have been shown to coalesce within cells, and can form and dissipate in response to biological and environmental signals. Many condensates have multiphase structure where different cellular components are sequestered into different layers, each with their own function. Here we discuss designing synthetic condensates to regulate biochemical reactions outside of cells. The synthetic condensates are formed via complex coacervation of oppositely charged polypeptides. Complex coacervation is a liquid-liquid phase separation that occurs via enthalpic charge interactions and entropic effects from the release of counterions and hydration water. We have created a library of different polypeptide sequences that incorporate various charge patterns, hydrophobic amino acids, and aromatic amino acids in the sequence. In addition, our sequences incorporate different chiral patterns of amino acids that bias the formation of liquid complex coacervates by preventing hydrogen bonding between the polypeptides know to form solid polyelectrolyte complexes. With this library we aim to explore the effects on the stability, encapsulation selectivity, and materials properties of the coacervate phase. Moreover, we mix coacervate phases with different interfacial tensions to create multiphase droplets and explore their function as synthetic condensates.

Using molecular engineering, nanoscale stabilization of polyelectrolyte complex formation can be achieved by coupling the polyelectrolyte to a neutral yet hydrophilic block, forming nanometer sized micelles with a polyelectrolyte complex core and a hydrophilic corona. Here we explore the potential of these micelles as delivery vehicles for charged molecules like proteins and nucleic acids. We characterize the polyelectrolyte complex micelles using scattering techniques and electron microscopy and explore the use of a thermosensitive neutral block on the structure and dynamics of the assemblies. Moreover, we evaluate the encapsulation of small and large molecules within the micelles as well as the role of phase behavior of the polyelectrolyte complex core on molecular exchange between micelles.

457b Disease in a Dish: Engineering Tissue Environment to Recreate Snapshots of Disease Progression

Srivatsan Kidambi

Engineering *in vitro* models that reproduce tissue microenvironment and mimic functions and responses of tissues that is more physiologically relevant represents a potential bridge to cover the gap between animal models and clinical studies. Our group aims to engineering in vitro models of tissues including cancer, liver, placenta, and brain in an effort to understand the role of the tissue microenvironment (physical attributes, cell-cell communication, and ligand density) on the underlying biology of healthy and diseased tissues. These platforms provide an ideal model to delineate the critical but unexplored areas of tissue microenvironment in which the cells reside. Specifically, we have developed matrix-based platforms that recreates various components of tissue microenvironment. These components include controlling the cell-cell interactions using patterned co-cultures and recreating the mechanical properties of tissues to provide a snapshot of physiologically relevant stages of the tissues in healthy and diseased state. Since tissue function is highly dependent on architecture, we have also used microfabrication methods, such as photolithography and molding, to regulate the architecture of these platforms. Using this strategy we have developed in situ models of breast cancer, liver, placenta, and brain. We have also developed clinically translatable nano-size liposome based drug delivery systems for encapsulating drugs ranging from high molecular weight proteins to small nucleic acids including miRNA, which regulates pathways important for disease progression and control. The technologies developed in our lab will have tremendous potential applications in the treatment of various diseases including liver fibrosis, cancer, preeclampsia, and development of several classes of therapeutic compounds (drugs, biologics).

457c Design of Targeted Antimicrobial Macromolecules

Christopher Alabi

Antimicrobials, including antibiotics, antivirals, and antifungals, are effective lifesaving therapeutics against pathogenic microorganisms. As the clinical pipeline of new antimicrobials runs dry, the emergence of resistance (bacterial and viral) against existing medicines poses a serious threat to public health. There is thus an urgent need for innovative, rationally designed molecular and nanotechnology platforms that can be rapidly developed to combat the elimination of resistant isolates. In the first part of this talk, I will describe our research effort toward creating antibody bactericide conjugates (ABCs). ABCs are targeted synthetic macromolecular prodrugs that actively target a pathogen of interest and release bactericidal macromolecules only in the presence of virulence or host factors. This mechanism of action, similar to that used in the field of antibody-drug conjugates (ADCs), should decrease the toxicity profile of the antibacterial agent while improving its potency at the infection site. ABCs can be considered a subset of ADCs, which have achieved clinical success for cancer treatment, with 11 FDA-approved ADCs available on the market and upwards of 80 in active clinical studies. I will present two unique ABCs and discuss their assembly, characterization, mechanism of action and antibacterial activity. In the second half of this talk, I will introduce the design of a fusion inhibitory lipopeptide platform with potent antiviral activity against several viruses, including Measles (MeV) and SARS-CoV-2. Viral infection of target cells occurs via the coordinated action of binding and fusion proteins. Peptides derived from the heptad repeat region of the viral fusion protein can interfere with the structural transition of the fusion protein, thus inhibiting infection at the entry stage. In a collaborative effort with the Porotto and Moscona research groups, our team has developed dimeric lipopeptide inhibitory ligands that self-assemble into serum stable nanoparticles with potent antiviral activity in cotton rats (MeV) and ferrets (SARS-CoV-2). I will discuss how self-assembly of the amphipathic lipopeptides enhances their biodistribution and half-life and contributes to enhanced *in vivo* efficacy. Proposed anchoring of the dimeric lipopeptide in the host cell membrane, interactions with the viral proteins, and lung retention will also be discussed.

509a Award Submission: Breast Cancer Specificity Assessment of Tumor Targeted Nano-, Encapsulated Manganese Oxide (NEMO) Particles

Celia Martinez de la Torre, Dhruvi Panchal, Kasey Freshwater, Margaret Bennewitz

Gadolinium chelates are the gold standard contrast agents for breast magnetic resonance imaging (MRI), but their constant MRI signal and lack of specificity results in high background and misdiagnosis. This lack of targeting makes it challenging to differentiate between benign and malignant breast masses, resulting in high false-positive rates of up to 25%. We have developed an enhanced contrast agent that utilizes tumor-targeted pH-sensitive manganese oxide (MnO) nanoparticles, which in physiological pH, produce low signal ("OFF"). After cancer cell uptake and digestion within low pH endosomes/lysosomes, the MnO dissociates into Mn^{2+}, turning the MRI signal "ON."

Nano-, Encapsulated Manganese Oxide (NEMO) particles were synthesized by the encapsulation of nanocrystalline MnO within poly(lactic-co-glycolic acid) (PLGA) and poly(ethylene glycol) (PEG) via single emulsion. The nanocrystals and nanoparticles were chemically and physically characterized by TEM, XRD, FTIR, SEM, and DLS. After characterizing the particles, a fluorescent tumor-targeting peptide against underglycosylated mucin-1 (uMUC-1) was attached through click chemistry using copper (I) as the catalyst. MUC-1 was chosen as a tumor target, as it is overexpressed and underglycosylated on breast cancer cells, but not on benign or healthy cells. The uMUC-1 conjugated NEMO particles were characterized with DLS to assess hydrodynamic size (~190 nm) and fluorescence to ensure attachment of the targeting peptide. Before evaluating the cellular uptake of the NEMO particles, different control and breast cancer cell lines were characterized for expression of MUC-1 and uMUC-1 via flow cytometry. HEK293T (control), BT20 (triple negative), and T47D (luminal A) were selected for particle specificity studies, as their expression varied from no MUC-1 or uMUC-1 (HEK293T) up to 85% uMUC-1+ cells (T47D).

Cells were exposed to targeted NEMO particles (5 μg of Mn/mL) for 2 hours, followed by an MRI scan of the cell pellet and collected media. The change of R_1 (ΔR_1) between the control group (only exposed to media) and the nanoparticles-exposed group was calculated post imaging. In the case of cell pellets, HEK293T had a larger ΔR_1 than BT20 or T47D (266% vs. 116% vs. 162%). This was not expected based on relative MUC-1 and uMUC-1 expression. However, when scanning media alone, the opposite results were observed where HEK293T had a lower ΔR_1 than BT20 and T47D (13% vs. 67% vs. 56%), meaning that higher amounts of Mn^{2+} were found in BT20 and T47D media. Increased Mn^{2+} in the media could indicate that cancer cells internalized NEMO particles faster, dissolved MnO, and then released it. Therefore, further characterization of the time course of cellular uptake was performed by exposing the cells to NEMO particles (1.625 μg of Mn/mL) for 0.5, 1, and 2 hr, and again calculating ΔR_1. Lower expression of uMUC-1 resulted in a smaller change in MRI signal at shorter time points. Already at 30 min, T47D had over 200% ΔR_1 due to its high uMUC-1 expression and thus fast NEMO particle uptake, while it took 2hrs for the other two cell lines to reach that level. These results showed the dependency of uMUC-1 expression on NEMO particle uptake and, consequently, MRI signal.

Lastly, T47D tumor-bearing mice were injected with targeted NEMO particles (7.5 mg/kg), and MRI scans were performed pre-injection (baseline) and 1, 2, 4, and 24hr post-injection to assess the variation in MRI signal in the tumor. After 1hr and 2hr, the MRI signal intersity increased by ~37%

compared to baseline, followed by a slight decrease to ~18% above baseline at 24hr. Additionally, through abdominal MRI scans, signal intensity was visualized in the liver and kidneys after NEMO particle injection, highlighting possible elimination via these routes. Future biodistribution studies will be performed to correlate MRI signal to Mn organ content. Overall, targeted NEMO particles caused an increase in T_1 MRI signal in breast cancer cells expressing uMUC-1+ in vitro and in vivo, and further study is warranted to compare its sensitivity and specificity to Gd-chelates.

509b Award Submission: Single-Walled Carbon Nanotube Based NIR Sensors for Measuring Enzymatic Depolymerization of Polyurethanes

Mei-Tsan Kuo, Jack Raffaelle, Nigel Reuel

Polyurethanes (PU) are synthetic polymers applied as coatings, adhesives, sealants and elastomers (CASE). They are designed to resist environmental factors, and thus the waste deposal of PU becomes an important issue. Enzymatic depolymerization is a promising, sustainable route to recycle polyurethanes waste; however, traditional characterization methods for determining the biodegradation rate of PU are time intensive and the sensitivities of these methods are low, leading to the poor compatibility to high throughput screening campaigns to find and optimize enzymes for processing polyurethanes. Thus, finding a new method to measure the biodegradation rate of polyurethane is needed. Herein, we propose the first single-walled carbon nanotube (SWCNT) based NIR fluorescence sensors for measuring the enzymatic depolymerization of Impranil® DLN, a commonly applied polyurethanes dispersion for qualitatively determining the depolymerization rate. Through wrapping the SWCNTs with cationic polymer, ionic bonds are formed between SWCNTs and Impranil, increasing the sensitivity of the sensor. Kinetic analysis and modeling show the sensor response follows a modified Langmuir adsorption model that both the Impranil and its depolymerization product would attribute to the surface coverage change of SWCNTs, leading to variation in the local dielectric environment and thus the fluorescence signal change of the SWCNTs. The presented sensor design provides a route for rapid measurement of enzyme activity towards Impranil and can potentially be applied to measure the enzymatic depolymerization rates of other kinds of conventional, fossil-fuel derived polymers. Such a tool will be useful in rapid prototyping of enzymes and microbes as a solution to remediating plastic waste.

Figure:
Upper panel: Schematic illustration of the synthesis method of the SWCNTs sensor
Lower panel: Typical time traces of the relative fluorescence signal change at 1000 nm of the SWCNT sensor. Time zero is set as the enzyme addition and the fluorescence signal intensity before adding enzyme is set as zero of the relative fluorescence signal intensity.

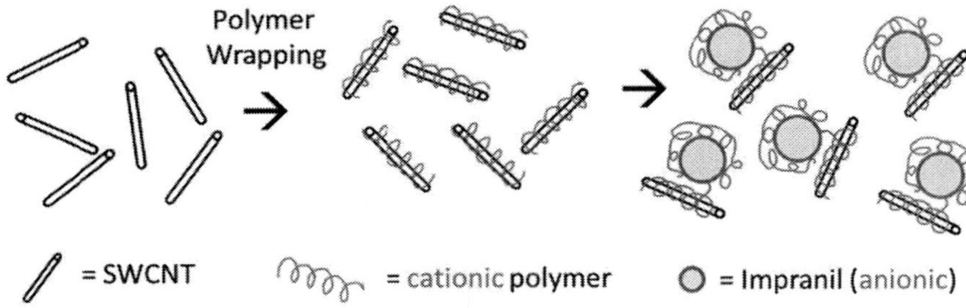
Polymer
Wrapping
= SWCNT
= cationic polymer
= Impranil (anionic)

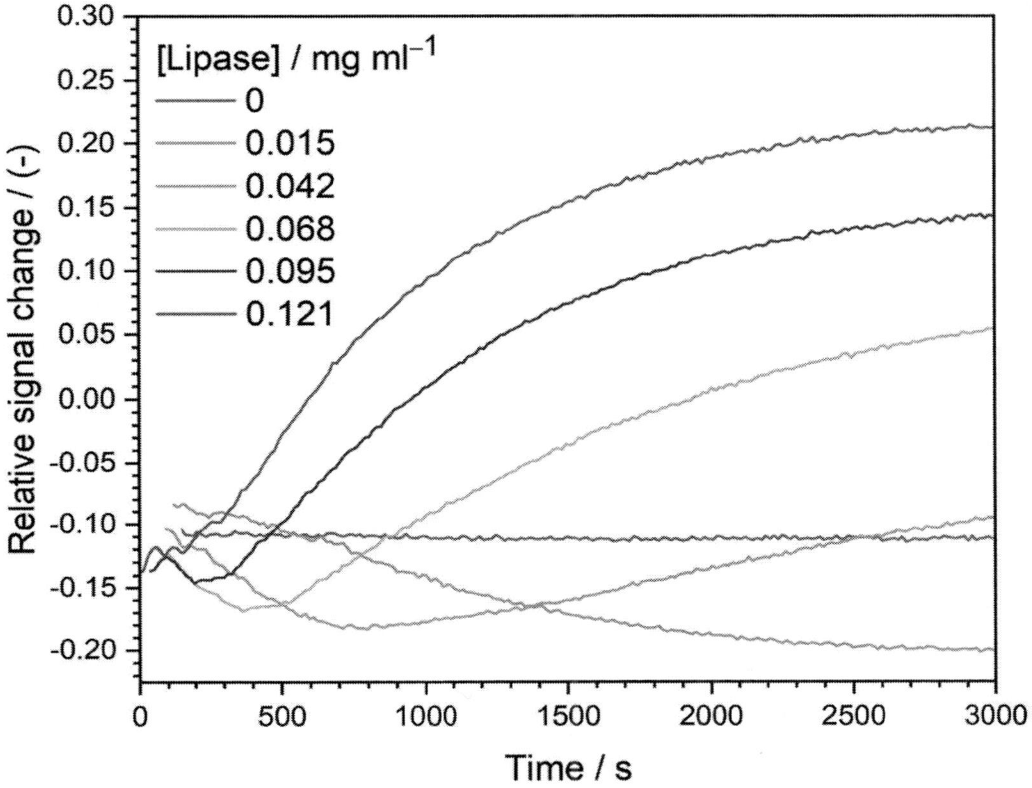
[Lipase] / mg ml^{-1}
0
0.015
0.042
0.068
0.095
0.121
Relative signal change / (-)
Time / s
0.30
0.25
0.20
0.15
0.10
0.05
0.00
-0.05
-0.10
-0.15
-0.20
0
500
1000
1500
2000
2500
3000

509c Antibody-Free Rapid Detection of Sars-Cov-2 Proteins Using Corona Phase Molecular Recognition to Accelerate Development Time

Xiaojia Jin, Sooyeon Cho, Xun Gong, Sungyun Yang, Jianqiao Leslie Cui, Michael Strano

To develop better analytical approaches for future global pandemics, it is widely recognized that new materials are necessary that enable molecular recognition and sensor assay development on a much faster scale than currently possible. Previously developed severe acute respiratory syndrome coronavirus 2 (SARS-CoV-2) point-of-care devices are based on the specific molecular recognition using subunit protein antibodies and protein receptors that selectively capture the viral proteins. However, these necessarily involve complex and lengthy development and processing times, and are notoriously prone to a loss of biological activity upon sensor immobilization and device interfacing, potentially limiting their use in applications at scale. Here, we report a new synthetic strategy for nanoparticle corona interfaces that enables the molecular recognition of SARS-CoV-2 proteins without any antibody and receptor design. Our nanosensor constructs consist of polyethylene glycol (PEG) - phospholipid heteropolymers adsorbed onto near infrared (nIR) fluorescent single-walled carbon nanotubes (SWCNT) that recognize the nucleocapsid (N) and spike (S) protein of SARS-CoV-2 using unique three-dimensional (3D) nanosensor interfaces. This results in rapid and label-free nIR fluorescence detection. This antibody-free nanosensor shows up-to 50% sensor responses within 5 min viral protein injections with limit of detection (LOD) of 48 fM and 350 pM for N and S protein, respectively. Finally, we demonstrate instrumentation based on a fiber optic platform that interfaces the advantages of antibody-free molecular recognition and biofluid compatibility in human saliva condition.

329a Photothermal Conversion Efficiency of Multi-Color Emissive Carbon Dots: Chemical and Thermal Analysis

Salar Balou, Aashish Priye

A series of highly stable, cost-effective, and environmental-friendly multi-color emissive carbon dots with systematically tailored chemical and physical properties were synthesized through a modified acid reagent engineered hydrothermal synthesis protocol. Detailed optical and thermal characterizations together with computational modeling revealed their remarkable photothermal (light-to-heat) conversion performance and the underlying mechanisms governing this observation were studied and discussed. Results showed a significantly higher light-to-heat conversion in red color emitting carbon dots compared to their blue, cyan, green, and yellow counterparts. Apparently, more acid treatment during the fabrication of carbon dots enhances the aromatization and introduces stronger electron donating functional groups (such as C-O-C, and C=O groups) into the lattice which gives rise to a more conjugated and oxidated carbon dot structure. In return, the HUMO and LUMO band gaps will get smaller which destabilizes the electrons in π-π^* orbitals and significantly increases the frequency of electron-phonon coupling in carbon dots. This phenomenon (similar to localized surface plasmon resonance) initiates a thermal vibration in the carbon dot lattice (the phonon-phonon coupling) which eventually dissipates heat into the surrounding media that we interpret as temperature increase in macroscopic scale. The highest thermal gradient observed was 11 °C/s in red color emitting carbon dots when illuminated with a blue laser (350 mW, continuous beam). Finally, these nanocarbons were employed as efficient photothermal nanomaterials to achieve ultrafast thermocycling for quantitative real-time PCR owing to their potential of implementing 30 cycles of 95 °C to 60 °C in less than 10 minutes. This work will set the first stage for bandgap engineering to utilization of carbon dots as efficient and effective nanomaterials for forward-looking photothermal applications and studies.

329b Unique Thermodynamic Co-Surfactant Equilibria of Single Walled Carbon Nanotubes for Fluorescent Biosensors

Aniruddha Kulkarni, Stephen Michel, Irene Chung, Claire Marc, Yang Zhao, Kirk J. Ziegler

Single walled carbon nanotubes (SWCNTs) have been extensively used for sensing as their unique optical properties are extremely sensitive to interactions with analytes. The ability to selectively modulate optical spectra in response to different biomolecules has been key in development of SWCNT-based optical biosensors. Semiconducting SWCNTs are a mixture of different chiralities and separation of monochiral species has been a major hurdle limiting their use to fluorescent sensors. Various surfactants have been used to form stable suspensions of SWCNTs in different solvents. Surfactant structure around the nanotube plays an important role in stability and separation of (n,m) species. The presence of thermodynamically stable co-surfactants states enables selective coating of species in solution. Our group has previously utilized co-surfactant solution approach to elute (6,5) species at a very specific ratio of sodium dodecyl sulphate (SDS)/sodium deoxycholate (DOC). This idea can be further extended to selectively target other chiralities and coat them in the nanotube solution.

Here, we utilize this thermodynamic equilibrium for selectively coating a mixture of semiconducting SWCNT chiralities such that we can achieve sensing on specific chiralities without separating them into monochiral fractions. As we add SWCNTs with different co-surfactants solutions, the reorganization of the surfactant shell around nanotubes results into solvatochromatic shifts which can be observed with optical spectroscopy. Difference in surfactant binding strength can be used to protect certain chiralities against further functionalization using addition of controlled ratios of co-surfactant solutions to nanotubes. Unprotected chiralities in solution were functionalized with different ligands depending upon the analyte biomolecule to be detected. We investigated different experimental approaches to chirality specific functionalization as well as the role of covalent and non-covalent functionalization on the nanotube surface to achieve the best biosensing response. Protected chiralities provide an in-situ reference which helps in quantifying analyte based on quenching of functionalized chiralities. This has led to development of a sensitive fluorescent SWCNT-based biosensor whose response can be tuned with the analyte concentration.

329c Estimation of the Structure of Confined Water between Hybrid Materials Using Convolutional Neural Networks (CNN)

Abhishek Sose, Fangxi Wang, Sanket Deshmukh

Two-dimensional (2D) materials such as graphene, boron nitride, and molybdenum disulfide find applications in a wide range of fields, including catalysis, sensing, energy storage, biology, and electronics. However, it is important to note that these materials are also highly dependent on factors such as surrounding temperature, pressure, and humidity. Our recent study investigated the structure of confined water between stacked 2D sheets of our novel layered hybrid materials using conventional analysis of molecular dynamics (MD) simulations trajectories. The principal factors like hydrophilicity and slit width between two 2D sheets significantly affect the structure of water confined. In this study, the spatial and chemical characteristics of sheets were integrated with deep learning models, *i.e.* Convolutional Neural Networks (CNN), to determine the precise structure of water around sheets. This study showcases the proof of concept of AI/Deep learning methods to model the structure of confined water around the hydrophilic/hydrophobic hybrid sheets. Specifically, the CNN model was trained on the three-dimensional structure of the hybrid material system in water to estimate the atomic density profile of oxygens in water and the infrared (IR) spectrum of water molecules. A 3D grid structure of water was allowed to convolute and extract the physical and chemical features that significantly affected the ordering of water around these hydrophilic/hydrophobic sheets. Four convolutional layers were fully connected, followed by five dense neural layers used to accurately train the model. Moreover, this surrogate model was used to accurately predict the structure of water around the sheets with their altered slit spacings and for the sheets replaced with their other counterparts.

329d High Performance 3D Printed Faradaic Supercapacitor Using Hybrid Nanocomposites of Reduced Graphene Oxide/MnO_x-Based Electrodes

Mahshid Mokhtarnejad, Erick L. Ribeiro, Dibyendu Mukherjee, Bamin Khomami

In this study hybrid nanocomposites (HNCs) based on manganese oxides (MnO_x) and reduced Graphene oxide (rGO) are synthesized as active electrodes for energy storage devices. Comprehensive structural characterizations demonstrate that the active material is composed of MnO_x/Mn_3O_4 nanorods and nanoparticles embedded in rGO nanosheets. The development of such novel structures is facilitated by the extreme synthesis conditions (high temperatures and pressures) of the liquid-confined plasma plume present in the Laser Ablation Synthesis in Solution (LASiS) technique. Specifically, functional characterizations demonstrate that the performance of the active layer is highly correlated with the MnO_x to rGO ratio and the morphology of MnO_x nanostructures in HNCs. To that end, active layer inks comprising HNC samples prepared under optimal laser ablation time windows, when interfaced with a percolated conductive network of electronic grade Graphene and carbon nanofibers (CNFs) mixture, indicate superior supercapacitance for functional electrodes fabricated *via* sequential inkjet printing of the substrate, current collector layer, active material layer, and gel polymer electrolyte layer. Electrochemical characterizations unequivocally reveal that the electrode with the LASiS synthesized MnO_x-rGO composite exhibits significantly higher specific capacitance compared to the ones produced with commercially available Mn_3O_4-Graphene NCs. Moreover, the galvanostatic charge-discharge (GCD) experiments with the LASiS synthesized HNCs show a significantly larger charge storage capacity (250 F/g) in comparison to NCs synthesized with commercially available Mn_3O_4-Graphene (158 F/g). Overall, this study has paved the way for use of LASiS-based synthesized functional material in combination with additive manufacturing techniques for all-printed electronics with superior performance.

329e Understanding Oligonucleotide Hybridization on Single-Walled Carbon Nanotube Corona Phases for Viral Sensing Applications

Jianqiao Leslie Cui, Xun Gong, Xiaojia Jin, Sooyeon Cho, Sungyun Yang, Michael Strano

Single-walled carbon nanotubes (SWCNTs) with tailored corona phases, or adsorbed molecules, have emerged as a promising material for sensing as the adsorption of an analyte can be specifically transduced as a modulation of their photoluminescence (PL). One such corona phase ripe for engineering is single-stranded DNA (ssDNA), where subsequent sequence specific hybridization can result in PL emission wavelength shifts. However, despite previous studies, the design principles for these ssDNA SWCNT constructs (ssDNA-SWCNTs) to result in both hybridization in conjunction with photophysical response is not well understood. In this work, we study the use of ssDNA-SWCNTs as a label-free approach to detect DNA and RNA oligonucleotides chosen specifically from regions of the SARS-CoV-2 genome for specific sensing. The complementary interactions of DNA and RNA oligonucleotides on the SWCNT corona results in the blue-shift of SWCNT PL emission wavelength with LOD of 10 nM following 1-hour incubation. Sensor specificity can be significantly improved, as compared to random-sequence controls by one order of magnitude, through the inclusion of additional nucleic acids as anchors in the adsorbed corona phase. $(GT)_{15}$ anchor resulted in greater wavelength shifts for both DNA and RNA targets compared to $(AT)_{15}$ and $(CT)_{15}$. Investigating hybridization incubation conditions show that bath sonication after incubation resulted in up to 22.1 percent enhancement in blue-shift. Additionally, co-incubation with 1 percent saliva preserved the blue-shift without compromising sensor specificity, while surfactant addition resulted in loss of sensor specificity. This work elucidates additional design principles for ssDNA-SWCNT-based hybridization nanosensors for targeting specific sequences toward the development of an adaptable point-of-care rapid test against nucleic acid targets.

329f Implications of Multiscale Graphene Interfaces on "Reverse" or "Inversed" Boiling

Seyed Alireza Rozati, Anju Gupta

Phase change heat transfer is crucial to various engineering applications. Some prominent examples include distillation reboiler, desalination, steam generators, high-powered electronic systems, and refrigeration. Boiling is a well-known process and has been studied extensively for decades. It is accompanied by phase change from liquid to vapor, subsequently, pool boiling involves the boiling of a stagnant liquid over a heated surface. It is desirable to enhance the boiling performance to enable higher heat dissipation to reduce equipment cost through surface engineering. For various manifestations, the pool boiling performance of a surface is dictated by higher critical heat fluxes and heat transfer coefficients with an increase in wall superheat. However, our graphene based surfaces have demonstrated an anomaly in the boiling curves with a decrease in wall superheat at higher critical heat fluxes which is reported as reverse boiling or "boiling inversion". We attribute this to various physical mechanisms such as the temperature gradients, separate vapor-liquid pathways, and impinging liquid jets that arise within the multiscale surface morphologies. This talk aims to summarizes various experimental and physical mechanisms of boiling inversion observed with multiscale graphene coatings. First, several combined powder metallurgy and additive manufacturing methods to form composite copper/graphene coatings will be presented along with their pool boiling performance. The second part of the talk will focus on correlating the tunable surface properties such as roughness, porosity, hydrophilicity, wickability and wicking rates, and with the boiling inversion mechanisms. This fundamental study will aid in overall thermal management of various high temperature equipment and devices.

NSEF Young Investigator Award - Research Acceleration in Nanoscience By Self-Driving Fluidic Labs

Milad Abolhasani

Despite the intriguing physicochemical properties and widespread applications of colloidal nanomaterials in energy and chemical technologies, their discovery, synthesis, and manufacturing are still based on Edisonian techniques. Existing nanomaterials development strategies using batch reactors with irreproducible and uncontrollable heat/mass transport rates very often fail to comprehensively explore the vast synthesis and processing universe of colloidal nanomaterials. Such resource-intensive experimental approaches result in a slow and expensive discovery and development of advanced functional nanomaterials (>10 years). Recent advances in reaction miniaturization and machine learning (ML)-assisted experimental space exploration techniques provide an exciting opportunity to reshape the discovery and manufacturing of colloidal nanomaterials and significantly accelerate the time-to-solution (from >10 years to <1 month). In this talk, I will present a 'self-driving fluidic lab (SDFL)' for autonomous discovery and manufacturing of emerging functional nanomaterials, with multi-step chemistries, through integration of flow chemistry, colloidal nanoscience, *in-situ* characterization, and ML. I will discuss how modularization of different nanomaterial synthesis and processing stages in tandem with a constantly evolving ML modeling and decision-making under uncertainty can enable a resource-efficient navigation through high dimensional experimental design spaces (>10^{20} possible experimental conditions). Example applications of SDFL for the accelerated development of clean energy nanomaterials will be presented to illustrate the potential of autonomous robotic experimentation in reducing synthetic route discovery timeframe by 100x compared to conventional techniques. Finally, I will present the unique reconfigurability aspect of SDFLs to close the scale gap in colloidal nanoscience through facile switching from parameter space exploration/exploitation to smart nanomanufacturing.

NSEF Forum Award - Nanomedicine: From High Tech to Global Health

Robert K. Prud'homme

Nanotechnology in drug delivery has a schizophrenic dichotomy of goals. One goal is to make drugs more bioavailable, which is normally associated with oral drug delivery. This bioavailability is associated with rapidly releasing drugs. The goal is achieved by making nanocarriers (NCs) with high surface-to-volume ratios, and with the drug in an amorphous state. The other goal is to encapsulate and deliver drugs to specific disease sites. This requires retaining the drug in the NC until targeted delivery is achieved. We will discuss examples of nanoparticle formulations based on our rapid micromixing platform – Flash NanoPrecipitation (FNP)– that address both of these goals. Sustained release of more hydrophilic drugs is achieved either by making insoluble ion pairs or through pro-drug synthesis. For oral delivery, enhanced dissolution of very hydrophobic drugs requires the high surface-to-volume ratio of NCs plus the drug being in a stable amorphous form in the NC core. Our research in oral nanoparticle formulations has been driven by funding from the Gates Foundation for low cost nanoparticle formulations. We demonstrate NC formation using lecithin, HPMC, and the corn protein, zein. The coupling of FNP to a spray drier enables a continuous, integrated, one step and scalable process for the production of powders for oral administration. The scalability of the platform is demonstrated from mg to kg scales. While FNP was initially developed for encapsulation of hydrophobic actives, soluble peptides and proteins are now the fastest growing segment of the pharmaceutical market. We will present a new *inverse* Flash NanoPrecipitation process (iFNP) which enables encapsulation of peptides and biologics at over 50% loading with 95% loading efficiency.

265a We Can Use Carbon to Decarbonize—and Get Hydrogen for Free

Matteo Pasquali

Constraints on CO2 emissions are confronting society with multiple massive challenges: creating new sources of clean energy beyond solar and wind; electrifying our transportation systems and using lighter weight materials; decarbonizing the industrial sector; and dealing with the economic fallout associated with shrinking the fossil hydrocarbon industry, which accounts for about 7% of the world economy. At first blush, the solutions to these dilemmas appear to be at odds with one another. In this lecture, I will show that there may be a way forward that positively influences all of these aims [1]: Our vision is a zero-emission future, where large scale technology is deployed to co-produce clean hydrogen and carbon materials that house, move, clothe, and feed people. To foster this vision, we have established the Carbon Hub [2] as a non-competitive partnership of industry, academia, government, and non-profit organizations.

1. Pasquali, C. Mesters, Opinion: We can use carbon to decarbonize—and get hydrogen for free. *Proc. Nat. Acad. Sci. USA*, **118**, e2112089118, 2021. https://doi.org/10.1073/pnas.2112089118
2. https://carbonhub.rice.edu/

265b Deconstructing Proton Transport through Atomically Thin Monolayer CVD Graphene Membranes

Piran Kidambi

Selective proton permeation through the atomically thin lattice of graphene and other 2D materials offers new possibilities for energy conversion/storage and novel separations. Practical applications necessitate scalable synthesis via approaches such as chemical vapor deposition (CVD) that inevitably introduce sub-nanometer defects, grain boundaries and wrinkles, and understanding their influence on H+ transport and selectivity for large-area membranes is imperative but remains elusive. Using electrically-driven transport of protons and other ions through monolayer CVD graphene we systematically deconstruct selective proton transport. Mitigation of large defects, wrinkles and tears via interfacial polymerization results in selective proton permeation rates comparable to state-of-the-art proton exchange membranes e.g. Nafion. Our work provides a new framework to assess H+ conductance and selectivity of large-area 2D membranes for practical applications.

265c Developing Carbon Nanomaterials Electrocatalysts for Active and Selective Carbon Dioxide to Target Products Conversion

Jingjie Wu

Electrocatalytic reduction of CO_2 into value-added chemicals or fuels is a promising technique for a carbon-neutral chemical process. The electrochemical reduction of CO_2 is a complicated process involving multiple protons coupled electron transfer, theoretically resulting in various products. Therefore, the major challenge in CO_2 reduction lies in manipulating the selectivity toward a specific product. However, the study on CO_2 reduction has not substantially advanced primarily because of the lack of fundamental understanding of the reaction mechanism and the challenge of discovering efficient and robust catalysts for the multi-electron transfer processes. Researchers have screened a wide range of metal materials for electrochemical CO_2 reduction and found only copper metals exhibit selectivity towards hydrocarbons and oxygenates with decent efficiencies, while most others favor the production of CO or $HCOO^-$. Here we present the development of carbon nanomaterials as an alternative to Cu for selective and high-rate electro-reduction of CO_2 into hydrocarbons and oxygenates. We will discuss the critical structural and electronic factors that govern the selectivity of carbon catalysts for the production of CO, CH_4, and C_2 products (e.g. C_2H_4 and C_2H_5OH). Three categories of carbon catalysts were developed based on the primary products of CO, CH_4, and C_2H_4 in our group. The first carbon catalyst featuring the metal-nitrogen-carbon structure exclusively catalyzes CO_2 electro-reduction into CO. The second catalyst, called functionalized graphene quantum dots (GQDs), can selectively reduce CO_2 into CH_4 with Faradaic efficiency (FE) of up to 70%. The third one, doped and H-passivated GQDs, yield $C_{\geq 2}$ products with FE of 70%. All catalysts show an industrial relevance current density of > 200 mA cm^{-2}. This study provides in-depth insights into developing high-performance carbon-based catalysts for electrochemical CO_2 reduction.

266a Combining Quantum Chemistry with Multiscale Atomistic Reactive Simulations to Develop New Nanomaterials for Energy Applications

William Goddard III

Critical energy applications for nanomaterials include **new electrocatalysts** for reduction of CO_2 to ethylene or ethanol, reduction of N_2 into NH_3, splitting H_2O into H_2 and O_2 gas, and also **new electrolytes and cathodes for use in batteries based on Li metal anodes**. To make faster progress in developing new nanomaterials to address these applications, we need improved methods for in silico optimization of nanomaterials suitable for tuning the properties at the interfaces of electrodes and electrolytes. We will discuss some of the new methods for electrocatalysis (Quantum Mechanics based Grand Canonical Potential Kinetics (GCP-K) methodology) and improved multiscale reactive force fields suitable for 10 nm (100,000 atoms) to 30 nm (million atoms) model systems and we will discuss applications to predicting Pourbaix diagrams and **turn-over-frequency (TOF) as a function of applied potential for electrocatalysis** and **formation of the electrode-electrolyte interface (EEI) for Li anode battery applications.**

The CO2RR material is based on work performed by the Liquid Sunlight Alliance, which is supported by the U.S. Department of Energy, Office of Science, Office of Basic Energy Sciences, Fuels from Sunlight Hub under Award Number DE-SC0021266 (Harry Atwater, PI). The OER and HER studies were supported by the US National Science Foundation (CBET-2005250, program mgr. Bob McCabe). The ORR studies were funded by the US Office of Naval research (ONR N00014-18-1-2155). The battery applications were funded by the Hong Kong Quantum AI Lab Ltd. in the frame of the InnoHK initiative (PI Guanhua Chen).

Formation of SEI layer (300 K, 1 ns) showing charge on Li

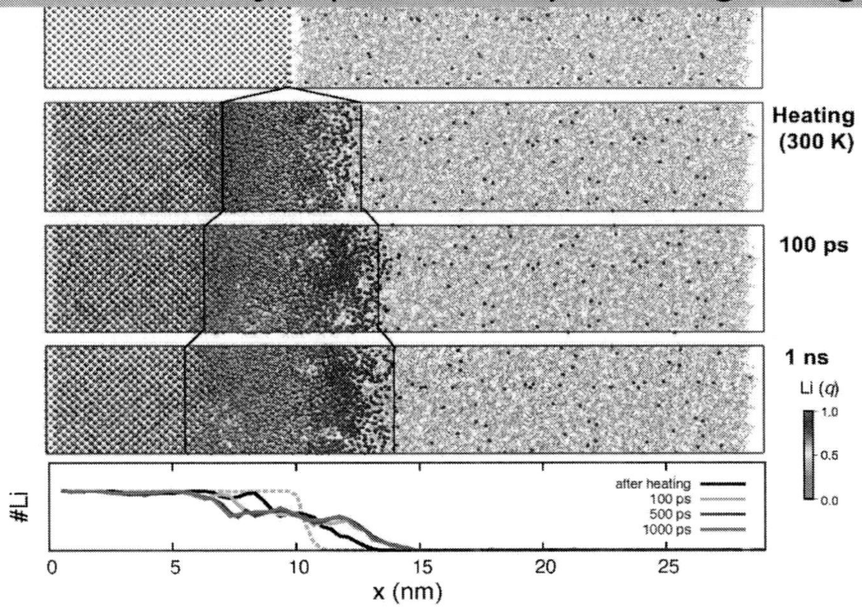

266b Tailoring Processes and Assembly of Polymer, Ceramic and Graphenic Materials for Electrochemical Energy Storage Applications

Yong Joo

Despite their promise to mitigate many problems and offer new opportunities in energy storage systems, nanomaterials have proven to be difficult to tailor and preserve their assembly throughout manufacturing processes when a scale-up is considered. My group has been working on understanding and modification of instability driven processes to devise scalable synthesis and manufacturing processes such as flow controlled continuous Taylor-Couette reactor, gas-assisted electrospinning and air-controlled electrospray which can also provide the tailored assembly of polymer, ceramic and graphenic materials at nano and micron scales. I will show several examples of using these processes to assemble various nanomaterials in the development of next generation battery materials, including: 1) thermally stable, non-flammable polymer/ceramic hybrid separator and high capacity silicon/graphene anode for high performance Li-ion batteries; 2) layered sulfur/graphene cathode for high rate capable Li-sulfur batteries, and 3) graphene coated electrodes for chemical flow batteries with high power density and high areal capacity.

266c Channel Engineering in Solid Polymer Electrolytes for Electrochemical Energy Storage and Conversion Devices

Paul Kohl

Ion conducting solid-polymer electrolytes (SPE) provide an enabling technology for safer and more compact low temperature fuel cells, hydrogen producing water electrolyzers, lithium ion batteries and flow batteries. The critical parameters of solid polymer electrolytes include ionic conductivity, ion selectivity, chemical resistance and dimensional stability in the presence of excess solvent or plasticizers. An SPE also provides an opportunity for achieving single ion conduction. Single ion conduction would lead to near unity transference number for the electroactive ion which would mitigate the major cause of concentration polarization and the resulting voltage drop at high current density. Conductivity is the product of the number of charge carriers times the mobility of each charge carrier. Although a relatively high density of ions within the SPE can be achieved, the optimum number is often limited by other factors including mechanical strength, solvent swelling and negative impact on ion mobility at high ion density. Ion mobility depends on microarchitecture within the SPE. Channel engineering can produce high mobility (and thus high conductivity) by forming efficient channels in terms of channel size and tortuosity. In this presentation, the SPE challenges in two systems (i) aqueous, alkaline fuel cell and electrolyzers and (ii) non-aqueous lithium ion batteries will be presented.

458a Light-Induced Chiral Metamaterial Printing

Ji-Young Kim, Connor McGlothin, Minjeong Cha, Emine Sumeyra Turali-Emre, Wonjin Choi, Nicholas Kotov

Imposing chiral bias during the synthesis of individual nanoparticles (NPs) and their assemblies enables to generate various 3D metamaterials with structural complexity. The chiral inorganic nanostructures with axial geometries aligned perpendicularly to the substrates are particularly the best for applications in meta-optics. However, due to the high thermodynamic penalties for out-of-plane alignment of nanocolloids, such geometry on surface/interface through self-assembly hitherto remains virtually unexplored. Therefore, most of manufacturing chiral inorganic superstructure on substrates now rely on the high-cost and time-consuming fabrication methods, which require either multistep lithography or depositions by instrumentation with high temperatures and/or low pressures. Here, we show that the strong optical activity of plasmonic materials under circularly polarized light (CPL) afford photon to matter chirality transfer, fabricating silver nanohelicoids (AgNH) on arbitrary substrates over the centimeter scale area in minutes. Illuminating CPL on substrate immersed in silver salt solutions induces the formation of NPs with 15–20 nm in diameter and their subsequent asymmetric deposition/assemblies. The handedness of the AgNH is altered from left to right when circular polarization of the incident photons changed from right to left, showing opposite polarity in their circular dichroism (CD) spectra. The wavelength of light source determined the size of the final superstructure as well as their CD band wavelength. Such dependence opens the path to printed metamaterials with 'on-the-fly' control of polarization rotation, showing tunability of their chiroptical activities in spectral range from UV to near infrared. These simple and versatile architected metamaterials can be used for extensive applications, including but not limited to photonic, optoelectronic, and electromechanical devices as well as enantioselective catalytic and sensing systems.

458b Effect of Pd Precursor Salt Type and Temperature on Synthesizing Five-Fold Pentagonal Pd Nanorods Using L-Ascorbic Acid in Segmented Millifluidic Flow Reactors (SMFRs)

Chamath Vindula Basnayake Pussepitiyalage, Shohreh Hemmati

L-ascorbic acid is a green reducing agent that has successfully demonstrated the ability to synthesize pentagonal Pd nanorods using Na_2PdCl_4 as the metal precursor in the segmented millifluidic flow reactors (SMFRs)[1]. Oxidative etching that is undesirable for formation of Pd nanorods is minimized in the SMFR by restricting presence of air in the reaction mixture thereby limiting air in the reaction mixture[2]. The presence of chloride ions originating from the Pd precursors such as Na_2PdCl_4 and $PdCl_2$ still causes a degree of oxidative etching that limits the synthesis of Pd nanorods. Alternative Pd precursors such as Na_2PdBr_4 and $Pd(NO_3)_2$ that do not contain chloride is used to improve the yields of synthesized Pd nanorods. Using Na_2PdBr_4 also had the additional benefit of supplying bromide ions that cap {100} facets of Pd decahedron seeds to promote 1-dimensional growth of nanostructures. The mechanism involved in the synthesis of Pd nanorods with each Pd precursor was identified using Transmission Electron Microscopy (TEM) and kinetic studies conducted via Ultraviolet–visible (UV-vis) spectroscopy. The effect of temperature variation on Pd nanostructure synthesis is examined and used to calculate the activation energy in synthesizing Pd nanorods with each Pd precursor. Reaction conditions are optimized to synthesize the highest possible yield of Pd nanorods. Finally, a filtration procedure is utilized to separate Pd nanostructures other than nanorods synthesized in the SMFR.

458c Controlled Polymer Nanoparticle Synthesis Using a Jet Mixing Reactor

Faiz Khan, Xiangming Gu, Jiaqi Luo, Nicholas Brunelli, Barbara E. Wyslouzil, Jessica Winter

Polymer nanoparticles have been widely used as carriers for encapsulated hydrophobic drug molecules. Such drug delivery nanocarriers provide therapeutic efficacy by overcoming biological barriers and improving circulation time in the body. Nanoparticle size and distribution are affected by process synthesis conditions and play a crucial role in determining therapeutic efficacy. Polymer nanoparticles synthesized through common nanoprecipitation techniques undergo rapid self-assembly in milliseconds. However, the mixing process in large batch reactors requires timescale on the order of minutes. This difference between the polymer self-assembly time and reactor mixing time results in a broad nanoparticle size distribution, with a high polydispersity. A polydisperse nanoparticle distribution compromises the uniformity of size-dependent properties, undesired for drug delivery applications. Hence, the goal is to achieve a reactor mixing time equivalent to the polymer self-assembly timescale.

It is difficult to achieve rapid mixing in large-scale batch reactor vessels. Thus, microreactor designs such as T-mixers, Y-mixers, and confined impinging jet mixers have been used for polymer nanoparticle synthesis. However, it has been reported that dual inlet microreactors designs fail to achieve effective mixing under asymmetric flow conditions, i.e., when the flow rates of two inlets are unequal. This significantly restricts exploration of reaction parameters, such as supersaturation and limits applicability in processes involving expensive reagents, such as drugs. Here, we decreased the reactor length scale to allow for a smaller reactor volume and thus, achieve rapid mixing while enabling operation under asymmetric flow conditions.

In this work, we present a three-inlet microreactor (i.e., jet mixing reactor, JMR) that comprises two opposing fluid streams that orthogonally impinge on a single, main fluid stream, developing vortexes and achieving rapid mixing in milliseconds. The JMR displays a cross-flow geometry and is the size of a dollar coin, with inlet diameters ranging from 0.5-1 mm. Mixing in the JMR was characterized using a competitive chemical reaction set, known as the Villermaux-Dushman reaction. Mixing dependence on inlet fluid stream flow rates, fluid viscosity, and cross-flow fluid ratios (R) was examined. Flow dynamics were also visualized using computational fluid dynamics simulations in COMSOL software. JMR performance was compared to that of conventional large batch vessels and a T-mixer using model nanoparticle (Polylactic acid-co-glycolic acid, PLGA, and Polybutylacrylate-polyacrylic acid, PBA-PAA) synthesis, with nanoparticle size and particle size distribution serving as an evaluation criterion.

Data showed that the JMR provides rapid mixing, with timescales between 0.5-100 milliseconds that depend on geometry, fluid velocity, and viscosity. These data were compiled to provide further insights into JMR scaling models. Different fluid mixing ratios (i.e., asymmetric flows) were examined in the JMR and T-mixer to explore the effect of supersaturation on polymer nanoparticle sizing. Because of momentum mismatch of two inlet streams, mixing in the T-mixer is not sufficient to achieve homogenous polymer nanoprecipitation kinetics under asymmetric flow ratios ($R = 10$, 25 and 50). However, the JMR mixes in milliseconds even under asymmetric flow conditions, providing smaller polymer nanoparticles at higher supersaturation conditions. PLGA nanoparticle studies were used for a comparison of batch versus JMR conditions. Rapid mixing in the JMR resulted in focused nanoparticle size distribution with a polydispersity of 0.16 versus polydispersity of 0.40 for batch synthesis. The mean nanoparticle size obtained through batch was ~2x times bigger compared to that obtained using the JMR. Thus, the JMR design holds promise to achieve smaller nanoparticle size and enhanced polydispersity control in rapid nanoprecipitation processes over batch stirred vessels and may allow greater operational flexibility over two-inlet microreactor systems for asymmetric flow conditions.

458d Continuous Millifluidic Synthesis of One-Dimensional Silver Nanostructures Using Tannic Acid As Reducing and Capping Agent: Parametric Study and Kinetic Behavior

Sina Kaabipour, Shohreh Hemmati

The synthesis of nanomaterials in a continuous setting such as in microfluidic devices has been gaining significant attention due to their capability to enhance mass and heat transfer and ultimately lead to a better control over nanostructures' size and morphology compared to that of the batch. However, microfluidic devices manufacturing is often costly and its use may be limited to small laboratory scale due to low throughput and as a result, may not be feasible for large-scale industrial production. Millifluidic reactors are able to maintain control over size and morphology, while also being cheaper to construct and are able to offer a higher throughput. This, however, depends on a number of factors such as the inner diameter of the reactor tubing/channel, tubing material, flowrate, residence time, and flow patterns arising from different geometrical configurations of the reactor. In this study, we were able to synthesize silver nanowires and nanorods in a helically-coiled millifluidic reactor in 96 minutes residence time by using merely tannic acid as both the reducing and capping agents. The synthesized silver nanostructures were characterized using Ultraviolet–visible Spectroscopy (UV-vis), Scanning Electron Microscopy (SEM), Transmission Electron Microscopy (TEM), and Energy Dispersive X-Ray (EDX). The silver nanowires synthesized in this process had a higher yield compared to that of the batch, which was done previously. We also investigated the effect of surfactant, light, various flow , and three different geometries of the reactor by changing the coil diameter. Additionally, the kinetic behavior was investigated by measuring the concentration of Ag^{+} ions in the millifluidic reactor at different residence times for three different coil diameters. The effect of coil diameter on the consumption of Ag^{+} ions, and subsequently the yield of silver nanowires, was analyzed and discussed. This study offers a versatile, inexpensive, and rapid approach towards the large-scale synthesis of silver nanowires in a continuous manner, without the need for using an external capping agent and hazardous reducing agents.

458e Sonochemical Method for High-Throughput Synthesis of Inorganic Nanostructures

Fabio Baum, Maria Politi, Cameron Tavakoli, Josiah S. Mace, Kiran Vaddi, Joshua Vasquez, Nadya Peek, Lilo Pozzo

The development of new materials is essential to help humankind to face many of its emergent challenges, such as the shift toward clean energy, the fight against pandemics, and granting access to clean water. To accelerate this process, new research initiatives, such as the Materials Acceleration Platforms (MAPs) are arising. The MAPs are a combination of tools for the discovery, optimization, and deployment of new materials. However, at the current state, MAPs still present significant challenges, such as inflexible designs (i. e. suitable only for specific reactions) and a high initial cost of implementation. In this context, our combination of open source robotic systems with sonochemical synthesis can achieve high-throughput experimentation with an implementation cost well below $10,000 US dollars. Our approach involves the use of liquid handling robots, OT-2 from Opentrons, and a custom-built motion platform with an ultrasonication tool, called 'sonication station'. The OT-2 liquid handling robot is an affordable tool capable of preparing hundreds of precursor solutions through careful liquid pipetting. The sonication station is a tool built onto an open-hardware 3D printing platform, Jubilee (https://github.com/machineagency/jubilee), that can sonicate samples through a sonication horn attached that allows an array of automated experiments. In a demonstration of the versatility of our platform, we synthesized four different classes of nanomaterials: cadmium selenide quantum dots, lead-free cesium bismuth bromide ($Cs_3Bi_2Br_9$) perovskite nanoparticles, upconverting nanoparticles (rare earth-doped $NaYF_4$), and exfoliated metal boride nanostructures (TiB_2 and AlB_2). To characterize the number of samples we generated, we also used high-throughput characterization techniques, such as commercially available microplate readers for UV-Vis spectroscopy and photoluminescence (PL) spectroscopy, robot-assisted high-throughput small-angle x-ray scattering (SAXS), and coating of sample arrays onto silicon wafers for x-ray diffraction (XRD). We believe our sonication platform can contribute to the scientific effort of acceleration in the discovery and optimization of new materials.

458f Extending the Diatom's Color Palette: Bio-Inspired, Colloid-Templated Structures for Vivid Coloration and Optical Sensing

Pavel Shapturenka, Isaac Zakaria, Helen Stute, Fabian Birkholz, Michael Gordon

Dielectric micro- and nanostructures have shown promise as exquisitely tunable and nearly lossless media for optical functions in both nature and technology. Herein, the biosilica valve structures of the centric marine diatom family, naturally used for cellular transport and UV-shielding, were computationally investigated with finite-difference time-domain (FDTD) methods as viable photonic elements and their optical behavior explored when comprised of a higher refractive index material (TiO_2). While sharp optical resonances exist within the native silica structure, they broaden and intensify within analogous TiO_2 nanostructures, leading to the possibility of filtering, high reflection, and wavelength-tunable design through proper selection of geometry. Non-dimensionalization of key structural parameters such as pore spacing and frustule thickness yielded a wavelength-invariant design space for intense (>95%) normal-incident reflection and transmission owing to the Fano resonances supported by the perforated, high-index thin film geometry. Furthermore, introducing disorder such as pore size polydispersity yielded varying degrees of modally specific, angle-independent scattering. To verify the computed results, a fabrication technique was developed involving area-scalable application of a colloidal patterning mask and standard nanofabrication. This process yielded non-iridescently colored surfaces across the entire visible wavelength range. This resonant filtering property was further implemented in a thin-film, porous Fabry-Perot refractive index sensor for liquids, which exhibited competitive sensitivity (~650 nm/RIU) and a striking color transition upon analyte infiltration. Such optical surfaces are area-scalable, substrate-agnostic, and may find further use as filters and waveguides in photovoltaics, passive displays, and optoelectronic devices.

458g Magnetically Aligned Metal-Organic Deposition (MOD) Ink Based Functional Surfaces with Enhanced Wettability

Seyed Alireza Rozati, Anju Gupta

We present enhanced phase change heat transfer achieved by magnetically aligned nickel precursor inks on copper substrates. Alignment was performed in single direction and grid orientation. Pool boiling studies were performed to obtain correlations between the heat flux, heat transfer coefficients and wall super heat. The effects of surface wettability and roughness on nucleate boiling heat transfer properties, and bubble dynamics is reported. Our studies yielded a critical heat flux (CHF) of 185.9 W/cm² and a heat transfer coefficient (HTC) 106.9 kW/m² °C for water on Ni grid patterned surface, representing an improvement of 49.9% in CHF and 105% in HTC compared to plain copper surface. The alignment of nickel followed by its sintering introduced nucleation characteristics that improved bubble dynamics. This is attributed to a) altered three-phase contact angle via chemically heterogeneity of Ni/Cu surface, and b) microlayer evaporation of confined fluids in the rough surfaces.

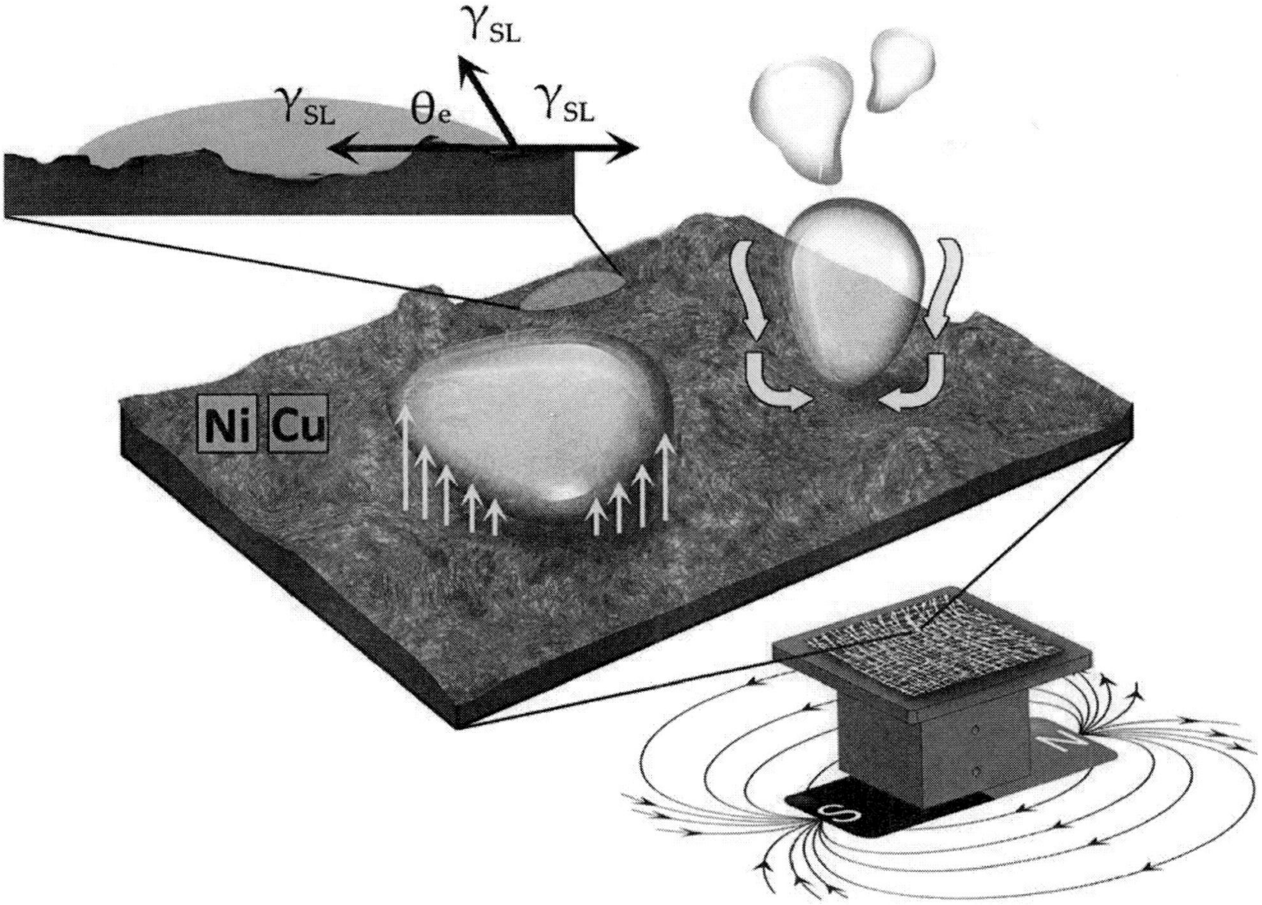

458h Assembly of Photoluminescent Silicon Nanocrystals with Reversible Covalent Bonds

Benjamin Stacy, Brian A. Korgel

Colloidal photoluminescent silicon (Si) nanocrystals are protected from oxidation by attaching organic ligands to the nanocrystal surface through hydrosilylation reactions of alkenes. Ligands with a wide variety of exposed functional groups can be tethered to the nanocrystals; thus, providing the opportunity for applications, such as contrast agents for biological imaging or as photosensitizers to drive molecularly-targeted redox reactions. Typically, Si nanocrystals are passivated by hydrosilylation reactions using high temperature or free radical initiators. We have shown that, in contrast to extended Si surfaces, hydrosilylation of Si nanocrystals smaller than about 5 nm in diameter can occur at room temperature, and that the yield of ligand-passivated Si nanocrystals is highly dependent on size. We also have used AIBN, a free radical initiator, to drive low temperature hydrosilylation of alkenes with aldehyde functional groups. The aldehyde has the ability to form reversible covalent bonds with hydrazide functional groups, which allows the creation of gels—networks of nanocrystals that can be assembled and disassembled depending on the reaction equilibrium. Here, we show the successful passivation of Si nanocrystals with alkenes with aldehyde functional groups at room temperature using AIBN. Hydrogen-terminated Si nanocrystals were first synthesized by thermal decomposition of hydrogen silsequioxane (HSQ), followed by HF etching. The aldehyde concentration on the nanocrystal surface could be adjusted by co-passivation with undecenoic acid and varying the mole ratio of aldehyde to undecenoic acid in the reaction mixture, as confirmed by characterization of the capping ligand layer composition using 1H NMR spectroscopy. Diffusion ordered spectroscopy (DOSY) NMR further confirmed that ligands with both acid and the aldehyde functional groups were bonded to the nanocrystal surface.

201a 2D Diamond Superstructures in Interlayer-Bonded Twisted Bilayer Graphene: Mechanical Response and Thermal Transport from Molecular-Dynamics Simulations

Mengxi Chen, Afnan Mostafa, Asanka Weerasinghe, Andre R. Muniz, Ashwin Ramasubramaniam, Dimitrios Maroudas

The thermomechanical properties of graphene derivatives and metamaterials fabricated through chemical functionalization and defect engineering of graphene sheets can be tuned by tailoring the structural features of these 2D materials [1]. Such exceptional properties make these materials particularly appealing for numerous technological applications. Here, we report a systematic computational study on the thermomechanical behavior of superstructures of 2D diamond nanocrystals embedded in interlayer-bonded twisted bilayer graphene (TBG) with commensurate bilayers. These superstructures are formed by patterned chemical functionalization, with hydrogenation chosen for this study, which induces interlayer covalent C-C bonding in TBG. The generated 2D diamond superstructures have the periodicity of the TBG Moiré pattern, with diamond nanocrystal sizes that can be varied by varying the extent and pattern of hydrogenation. These structures are fully defined by specifying parameters that include the commensurate bilayer's twist angle, the interlayer bond pattern and density, and the concentration of sp^3-bonded C atoms that serves as a metric of the nanodiamond fraction in the superstructures [2].

We report results for the mechanical behavior of these graphene-diamond nanocomposite superstructures based on molecular-dynamics (MD) simulations. We have conducted uniaxial straining tests and report the dependence on the diamond fraction of the elastic properties, ultimate tensile strength, and fracture strain of the 2D diamond superstructures, demonstrating their remarkable mechanical strength. We find that a brittle-to-ductile transition occurs in these superstructures with increasing the 2D diamond concentration beyond a critical level [2]. The underlying ductile fracture mechanism is mediated by void formation, growth, and coalescence, in contrast to the typical brittle fracture of graphene. Furthermore, we have analyzed the response of these superstructures to nanoindentation loading [3] and to shear straining. We find that superstructures with a less-than-critical 2D diamond concentration exhibit a strongly nonlinear inelastic response to indentation up to the onset of fracture mediated by a non-dissipative and non-recoverable stiffening effect that results in large hysteresis loops in indentation loading/unloading cycles. We also demonstrate the impact of increasing the 2D diamond fraction in the superstructures on the shear modulus [1] and shear strength of these 2D nanocomposite materials.

Finally, we report results for the lattice thermal conductivity of these superstructures based on equilibrium MD simulations. We compute the dependence of the thermal conductivity on the 2D diamond fraction in the nanocomposite superstructures and provide a comprehensive theoretical interpretation of the simulation results based on an effective medium model and detailed analysis of the internal displacement fields in the superstructures due to strained C-C bonds in these 2D materials. The findings of this thermal transport analysis have important implications for the suitability of these 2D materials for thermal management applications.

1. D. Maroudas, A. R. Muniz, and A. Ramasubramaniam, Molecular Simulation **45**, 1173–1202 (2019).
2. M. Chen, A. R. Muniz, and D. Maroudas, ACS Applied Materials andamp; Interfaces **10**, 28898–28908 (2018).
3. M. Chen, A. Weerasinghe, A. R. Muniz, A. Ramasubramaniam, and D. Maroudas, ACS Applied Nano Materials **4**, 8611–8625 (2021).

201b Carbon Nanotube Length Reduction from Planetary Ball Milling

Mason Rhue, Brian Grady

Carbon nanotubes show great evidence as additives in polymer composites to boost desired mechanical and electrical properties. However, the cost of carbon nanotubes currently limits their viability in industry as more traditional carbon forms like carbon black and carbon fiber yield similar boosted properties with lower costs. One strategy to lower the cost of carbon nanotubes is to increase yield by increasing nanotube length grown on conventional supported catalysts to upwards of 1 mm, eliminating the need to separate carbon nanotube product from catalyst and support material. With this comes the challenge of dealing with very long carbon nanotubes which can cause a variety of processing issues when compounded into a polymer. In this study, both very long and more standard length commercial carbon nanotubes were broken in a planetary ball mill. The length reduction of these carbon nanotubes was quantified using SEM imaging and ImageJ/FIJI image processing software. The Burgio-Rojac model for planetary ball milling was used to correlate the single impact energy and cumulative energy of milling to carbon nanotube length. It was found that the single impact energy varied for each nanotube type, likely due to differences in nanotube diameter. It was also found that the length reduction of the nanotubes for increasing cumulative energies was independent of mass of nanotubes in the mill and initial packed density of the nanotubes. An exponential function that models the length reduction of nanotubes based on the cumulative energy of milling was developed.

201c Porosity and Crystallinity Dynamics of Carbon Black during Internal and Surface Oxidation

Georgios Kelesidis, Nicola Rossi, Sotiris E. Pratsinis

Oxidation of carbonaceous materials (e.g. coal, biomass, soot and carbon black, CB) determines their emitted mass concentration (Kelesidis and Pratsinis, 2021a), production rate and performance in various applications (Kelesidis and Pratsinis, 2021b). Here, the structure of CB is investigated during its oxidation by O_2 at 550 and 800 °C by thermogravimetric analysis, microscopy, N_2 adsorption and X-ray diffraction (XRD). That way, the dynamics of its porosity and crystallinity during surface and internal oxidation are elucidated, for the first time to the best of our knowledge. At 800 °C, oxidation takes place largely on the CB surface. This increases the CB specific surface area, *SSA*, from 240 to about 453 m^2/g (Fig. 1), closely following the classic shrinking particle model and hardly affects porosity and crystallinity of CB. At 550 °C, O_2 diffuses and reacts with the bulk of CB particles forming pores with radius smaller than 2 nm after 25 and 50 % conversion. Fractal analysis of these pores and XRD reveal that internal oxidation reduces the interlayer spacing of the constituent crystallites of CB. This compacts its pore network and increases the pore fractal dimension, D_{fp}, from 2.43 up to about 2.7, in good agreement with the D_{fp} measured during diesel soot oxidation. At higher conversions, the small pores fuse resulting in hollow particles with SSA = 939 m^2/g. These structures are similar to those obtained by low temperature oxidation of biodiesel and diesel soot, as well as to oxygen-treated CBs (e.g. Ketjenblack) used in Li-ion batteries. So, the low temperature oxidation dynamics of CB porosity and crystallinity quantified in detail here can assist the derivation of robust oxidation rates (Kelesidis and Pratsinis, 2019), as well as the design of highly porous carbonaceous nanomaterials for filters and battery electrodes.

References:

Kelesidis, G. A., and Pratsinis, S. E. 2019. Estimating the internal and surface oxidation of soot agglomerates. Combust. Flame 209:493-499, doi.org/10.1016/j.combustflame.2019.08.001.

Kelesidis, G. A., and Pratsinis, S. E. 2021. Determination of the volume fraction of soot accounting for its composition and morphology. Proc. Combust. Inst. 38:1189-1196, doi.org/10.1016/j.proci.2020.07.055.

Kelesidis, G. A., and Pratsinis, S. E. 2021. A perspective on gas-phase synthesis of nanomaterials: Process design, impact and outlook. Chem. Eng. J. 421:129884-1-7, doi.org/10.1016/j.cej.2021.129884.

Figure 1. Microscopy images and *SSA* of (a) as-received and (b-g) oxidized CB at 550 (b-d) and 800 °C (e-g) after 25 (b, e), 50 (c, f) and 75 % (d, g) conversion.

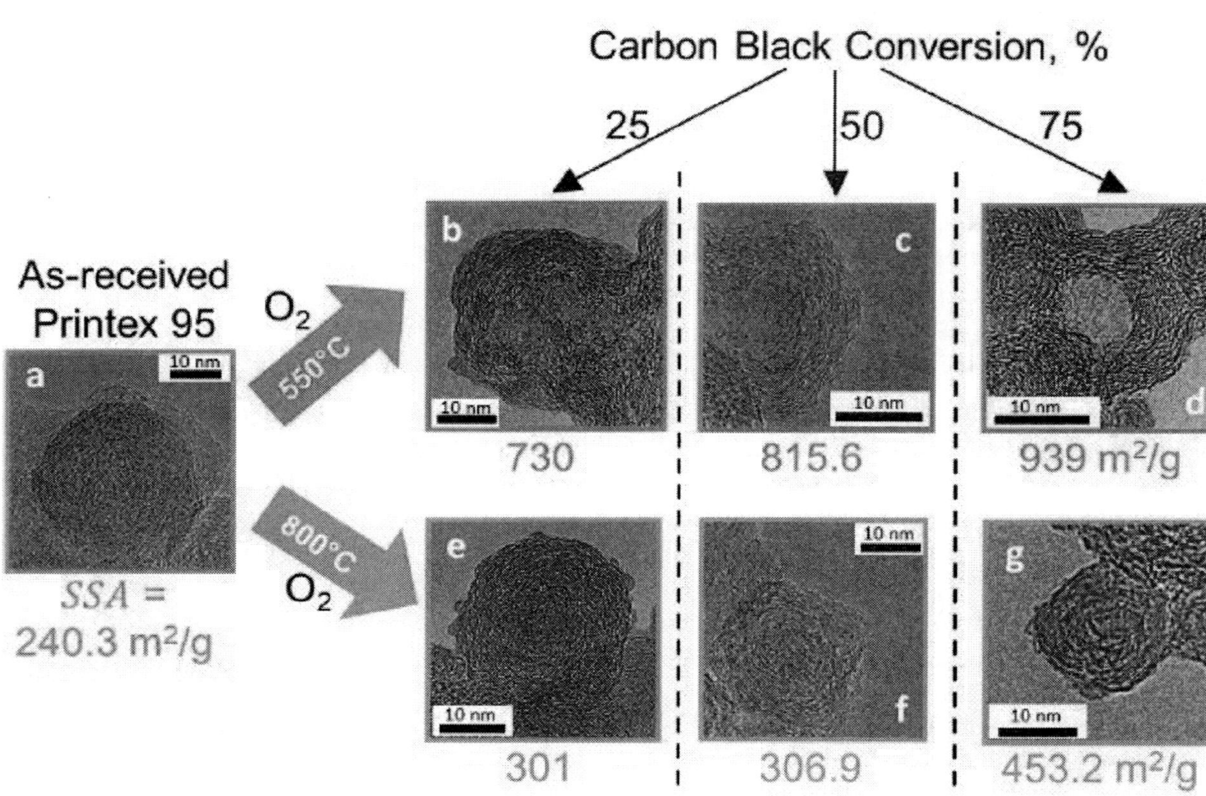
Carbon Black Conversion, %
25
50
75
As-received
Printex 95
O2
550°C
800°C
O2
a
b
c
d
e
f
g
10 nm
730
815.6
939 m²/g
301
306.9
453.2 m²/g
SSA =
240.3 m²/g

201d Automated and High Precision Measurement of Carbon Nanotube (n,m) Dependent Extraction Conditions in Aqueous Two Phase Extraction

Christopher Sims, Jeffrey Fagan

Aqueous two-polymer phase extraction is a sensitive method for separating specific (n,m) structures of single-walled carbon nanotubes (SWCNTs). For surfactant-dispersed SWCNTs, the selecting mechanism for top versus bottom phase extraction in a given separation depends sensitively on the type and chemistry of the adsorbed surfactant layer on each (n,m) species. In practice, two surfactants, sodium dodecyl sulfate (SDS) and sodium deoxycholate (DOC), are generally competed for adsorption on the SWCNT-solution interface, with SDS-coated and DOC-coated (n,m)s partitioning to opposite polymer phases. Recently we demonstrated that the nature of the adsorbed interfacial surfactant layer on each (n,m) species could be determined using near-infrared fluorescence spectroscopy as a function of the solution concentrations of both surfactants. This allows determination of the critical surfactant concentration necessary to change the phase of extraction *without* actually conducting a separation, and enabled initial determination of such values (PCCC values) for several (n,m) species and determination of surfactant type, surfactant structure, and solution condition effects on the ATPE separation. In this presentation I will describe the extension of the PCCC measurement to greater precision and scope through the automated measurement of gradients in surfactant compositions, including measurements on separability of enantiomers for a greater range of (n,m) SWCNT species and overall impacts to control of (n,m) selection through separation design.

201e Divalent Metal Cation Optical Sensing Using Single-Walled Carbon Nanotube Corona Phase Molecular Recognition

Xun Gong, Sooyeon Cho, Michael Strano

Colloidal single-walled carbon nanotubes (SWCNTs) offer a promising platform for the nanoscale engineering of molecular recognition. Optical sensors have been recently designed through the modification of non-covalent corona phases (CPs) of SWCNTs through a phenomenon known as corona phase recognition (CoPhMoRe). In CoPhMoRe constructs, DNA CPs are of great interest due to the breadth of the design space and our ability to control these molecules with sequence specificity at scale. Utilizing these constructs for metal ion sensing is a natural extension of this technology due to DNA's well-known coordination chemistry. Additionally, understanding metal ion interactions of these constructs allows for improved sensor design for use in complex aqueous environments. In this work, we study the interactions between a panel of 9 dilute divalent metal cations and 35 DNA CPs under the most controlled experimental conditions for SWCNT optical sensing to date. We found that best practices for the study of colloidal SWCNT analyte responses involve mitigating the effects of ionic strength, dilution kinetics, laser power and analyte response kinetics. We also discover that SWCNT with DNA CPs generally offers two unique sensing states at pH 6 and 8. The combined set of sensors in this work allowed for the differentiation of Hg^{2+}, Pb^{2+}, Cr^{2+} and Mn^{2+}. Finally, we implemented Hg^{2+} sensing in the context of portable detection of within fish tissue extract, demonstrating nanomolar level detection.

563a Probing Brain Structure-Function Relationships in Neurodegeneration Using Organotypic Whole-Hemisphere Slice Models and Multiple Particle Tracking Technology

Brendan Butler, Elizabeth Nance

Background: Aging is a common denominator in neurodegenerative pathogenesis, however the mechanisms by which healthy aging deviates towards a pathological state remain unclear. We hypothesize that the structural integrity of brain becomes compromised due to metabolic stress on neurons, rendering the brain less capable of maintaining healthy states and becoming more susceptible to pathologies such as plaque buildup or Lewy body aggregation. The use of nanoparticles with real-time tracking technology has increased access to structural information on live brain tissue – prior work using nanoparticle tracking techniques has identified design parameters that enable penetration in the brain microenvironment and detection of microstructural changes associated with different brain ages (1,2).

In this work, develop a neurodegeneration model in organotypic whole hemisphere (OWH) brain slice tissue. OWH brain slices are robust, stimuli-responsive platforms for recapitulating *in vivo* functional states with high throughput and spatiotemporal resolution (3,4). To model pathological processes in neurodegeneration in OWH slices, we induce chronic energy depletion in brain cells. Metabolic decline, driven by mitochondrial abnormality, is a classical hallmark of pathological aging and onset of neurodegeneration (5,6). Continuous, low-grade exposure to the neurotoxin rotenone (ROT) inhibits Complex I of the mitochondrial respiratory chain at intercellular concentrations above 20 nM (7). 50 nM ROT exposure has been shown to result in sufficient damage to dendritic spines and reduce neuronal connectivity (8). The upper limit of ROT exposure, without compromising viability, is reported to be 20 μM (9). We use the ROT-exposed OWH brain slice model in combination with multiple particle tracking, confocal microscopy, and molecular biology techniques to probe micro and macroscopic changes to native extracellular architecture and cellular function.

Methods: We develop the OWH platform to: 1) model the development of pathological processes that contribute to neurodegeneration, 2) characterize regional hallmarks of neurodegeneration at the inter-, intra-, and extra-cellular level, and 3) correlate these data with microstructural information. Based on the previously reported limits of ROT exposure, ROT doses in this study range from 50 nM to 10 μM. ROT is diluted in DMSO at 10 mM and further diluted into aqueous culture media at 10 μM (DMSO $< 0.01\%$ v/v). Slices are cultured 4 days in normal media – this is based on evidence that viability loss from the slicing procedure recovers after 4 days *in vitro* (4DIV) for slices from postnatal day 10 (P10) rats (10). This postnatal age is well-suited to study chronic effects of ROT exposure as P10 slices maintain viability for up to three weeks post extraction. At 4DIV, media is replaced with ROT media and changed every day for 6 days of ROT exposure. We assume that continuous ROT exposure establishes a uniform concentration profile of ROT in cells. Healthy control slices (NC) continue to receive normal media and are provided fresh media at the same time points as ROT-exposed slices.

Structure-function relationships are studied by correlating changes in tissue microstructure with effects of ROT exposure. Fluorescent bioinert nanoparticles are incubated in OWH slices and tracked dynamically using live-video epifluorescence microscopy. Nanoparticle trajectories are segmented using ImageJ TrackMate (11) and fed into a lab-developed Python package *diff_classifier* to compute trajectory mean-squared displacements (MSD) and geometric features (12). Particle trajectories can be classified as Brownian and non-Brownian, for which Brownian MSD values directly interface with governing transport equations to yield values for local effective diffusivity and viscosity. We perform multiple particle tracking (MPT) analysis in multiple brain regions to study the regional effect of ROT-exposure on the extracellular space.

We further assess the effects of ROT-exposure at the organelle, cellular, and extracellular levels. We isolate synaptosomic mitochondria from NC and ROT slices and assayed for activity (Sigma-Aldrich, MAK259). Slices are homogenized, decanted, and centrifuged in a mitochondrial isolation buffer to collect crude mitochondrial pellet, further separated into synaptosomic and non-synaptosomic mitochondria by discontinuous Percoll gradient in MIB (13). At the cellular level, we examine whole-slice viability and the functional health of neurons by region. Whole-slice viability is assessed by a lactate dehydrogenase (LDH) assay, for which slice culture supernatants are assayed at each time point from NC and ROT-exposed slices. The cumulative percent of damaged cells is computed relative to a maximum death control corresponding to the highest dose (10 μM) and longest exposure (6d) OWH slices treated with Triton-X detergent. We evaluate neuronal health in the following regions of interest (ROIs): cortex (CTX), striatum (STR), hippocampus (HP), and thalamus (TH). Neuronal death by ROI is quantified by co-localization of propidium iodide signal (a marker of dying or dead cells) with neurons, imaged using confocal microscopy. Slices are also stained with the neurodegenerative marker FluoroJade-C (FJC) and microtubule marker MAP-2, both of which assess structural integrity of neurons in response to ROT (8, 14). At the extracellular level, we quantify matrix degradation by extracting proteoglycans and using liquid-chromatography with mass-spectrometry (LC/MS) to survey reductions in average molecular weight of components such as hyaluronic acid and chondroitin sulfate. Collectively, this set of studies interfaces structural information gathered by nanoparticle probes with multi-level biological response to metabolic decline.

Results: We observe tunable responses to ROT exposure in our OWH model. Assaying for whole-slice viability demonstrated dose- and time- dependent responses to ROT exposure. The increase in cumulative cytotoxicity, relative to NC slices, was proportional to the dosing level and this relationship was maintained over 6 days of continuous ROT exposure for each dose. At 6 days of ROT exposure, slices exposed to 50 nM ROT experience a 25% increase in cumulative cytotoxicity. The most severe exposure (highest dose, longest time) reached a cumulative cytotoxicity of 86%. We observed a proportional increase in the expression of matrix-metalloprotease 9 (MMP-9) in response to ROT exposure, providing preliminary evidence that ROT exposure drives structural modifications while simultaneously inflicting cellular damage. Regionally, MMP-9 expression (mean fluorescence signal, n=10 images per group) increased 5-fold in the cortex and 3.5-fold in the striatum following 6 days of 1 μM ROT exposure, compared to 1.7- and 1.8-fold increases in NC slices. Diffusion data from MPT shows a change in nanoparticle diffusion ability dependent on both ROT exposure time, dose, and region. Diffusivity profiles for acute exposure (24h) contrast with long exposure (6-day) results. Nanoparticles were tracked at 2h, 6h, and 24h for 50 nM and 10 μM exposure. 50 nM ROT showed similar acute decay profiles for effective diffusivity (D_{eff}, $\mu m^2/s$) in both the cortex and striatum, decreasing from 0.55 to 0.14 (CTX) and 0.46 to 0.14 (STR) over 24h. For 10 μM exposure, both regions experienced a sharp drop in diffusivity within 2h (CTX, 0.55 to 0.13; STR, 0.46 to 0.03) followed by an increase to 71% (CTX) and 43% (STR) of the initial NC D_{eff} within 24h. Interestingly, for longer exposure, response diverged by region and dose. Following 6 days of ROT exposure, NC effective diffusivity in the cortex increased from 0.55 (NC) to 0.70 (50 nM), 0.60 (5 μM), and 0.59 (10 μM), relative to a 6d NC diffusivity of 0.14. In the striatum, effective diffusivity responded differently, increasing from 0.46 (NC) to 0.67 (50 nM), and decreasing to 0.08 (5 μM) and 0.42 (10 μM), relative to a 6d NC diffusivity of 0.09. The extent to which ROT-driven mitochondrial malfunction contributes to neuronal abnormality and tissue structural modifications will continue to be revealed by the intercellular and cellular analyses described above.

Conclusions: The ROT-exposure OWH slice model provides a tunable platform to survey the effects of brain cell metabolic stress on short- and long-term neuronal health and tissue structural integrity. We precisely tune and conserve whole-slice cellular response by dose and exposure time, represented by LDH assay data. Observable region-dependent changes in surrogate features of tissue microstructure enable further investigation into the time- and severity- dependent impacts of metabolic stress on neuronal health and brain microstructure. Our ROT-OWH model can equally screen recovery from exposure to explore the questions of irreversibility and how permutable disease states mediate tissue resiliency.

References:

1. Nance et. al. (2012). 1126/scitranslmed.3003594

2. McKenna et. al. (2021). 10.1021/acsnano.1c00394

3. Joseph et. al. (2020). 1002/btm2.10175

4. Wood et. al. (2020). 10.1002/btm2.10265

5. Lau et. al. (2013). 10.1038/nrn3550

6. Grimm andamp; Eckert. (2017). 1111/jnc.14037

7. Higgins et. al. (1996). 10.1523/jneurosci.16-12-03807.1996

8. Testa et. al. (2005). 1016/j.molbrainres.2004.11.007

9. Pakrashi et. al. (2020). 10.1007/s11064-020-03061-8

10. McKenna et. al. (2022). In review

11. Schneider et. al. (2012). 1038/nmeth.2089

12. Curtis et. al. (2019). 10.21105/joss.00989

13. Lampl et. al. (2015). 3791/52076

14. Lapointe et. al. (2004). 10.1096/fj.03-0677fje

563b Long-Term Dose-Controllable Drug Delivery Implant

Yoonjee Park, Xingyu He, Zheng Yuan

Purpose: The overall goal of this research is to advance treatment of chronic diseases by addressing current challenges in long-term dose-controllable drug delivery, with a focus on chronic posterior eye diseases. Repetitive intravitreal injections of drug or intravitreal injections of sustained-release corticosteroid implants (Ozurdex, Retisert, and Iluvien) are the current standard of care for the chronic diseases. However, these methods involve complications due to the high initial dose at the time of injection, known as burst release.
In this study, we have developed a size-exclusive nanoporous biodegradable poly lactic glycolic acid (PLGA) capsule for dosage-controllable drug delivery implants to avoid the burst release and control the dose only when needed.

Methods: We have developed a biodegradable PLGA implant enclosing light-activated liposomal drug (methotrexate). The PLGA implant has nano-sized pores where methotrexate (MTX) was released through when activated by laser, leaving the liposomal drug in the capsule. We have optimized the porosity and the pore size for MTX release kinetics, and tested the stability and the safety for 6 months in vivo rabbit eyes. We also irradiated near infrared laser (NIR, 1064 nm) through the lens of rabbit eyes to release MTX and fluorescence dye for visualization. The MTX release by laser was quantified based on in vivo imaging.

Results: We created a pore size less than 5 nm to selectively release drug molecules only upon laser irradiation leaving the liposomal drug inside the capsule ($p<0.05$). We observed the nanopore size increased during degradation over 6 months in physiological conditions via scanning electron microscopy.
Shrinkage of the implant structure, observed on Day 180, was attributed to the degradation process. No adverse event due to the implant was observed on the retina via fundus, ultrasound exam and histology, during the 6 months. The results also indicate that the liposomal drug released by degradation did not cause any damage to the retina. The implant did not seem to interfere with vision of the rabbits, based on their behavior. The location and position of the implant was not significantly changed.

Lastly, the dose released in vivo by laser followed the first-order kinetics and daily/weekly dose was clinically relevant (~50 ug per week). Histology and optical and ultrasound imaging data showed no abnormality in the eyes received implant treatment, suggesting that the drug delivery system was safe to the retina.

Conclusions: In this study, we successfully showed effective drug release from a nanoporous PLGA implant using pulsed NIR laser irradiation both in vitro and in vivo. The drug delivery system was safe and stable against leakage for 6 months. The light-activated drug delivery system that we developed will provide tightly-controlled release whenever it is necessary without burst release. Thus, the drug delivery system could be potentially used for long-term posterior eye disease treatment.

563c Reconfiguration, Manipulation, and Control of Higher Order Dynamic DNA Origami Assemblies

Anjelica Kucinic, Teng Teng, Dylan Roderick, Ratnasingham Sooryakumar, Carlos E. Castro

Over the last few decades DNA has garnered interest as a material for bottom-up- self-assembly of new devices and materials for various biological, mechanical, and chemical applications. DNA nanotechnology, specifically DNA origami, provides a platform that leverages Watson-Crick complementary base-pairing to design nanoscale devices with a pre-defined geometry capable of precise motion and controlled mechanical properties. The self-assembly process and tunable mechanical properties of DNA nanostructure components provides a framework to design dynamic nano devices capable of higher order assembly and reconfiguration. Methods have also been developed to reconfigure individual nanodevices into many different states, but to date the actuation of higher order assemblies is limited to simple reconfiguration like opening or closing of hinge joints. The most common approach to actuate reconfiguration in these devices is toe-hold mediated strand displacement, which relies on the formation or dissociation of specific DNA interactions that control the device conformation. While strand displacement is a robust method of actuation, limitations include the slow response time (minutes or greater) and the need to introduce strands into solution. To overcome these limitations recent efforts have established methods that leverage externally applied fields to manipulate structures. For example, recent efforts developed an approach to integrate magnetic microbeads for direct real-time manipulation of devices with sub-second time scales, tunable applied forces, and precise spatial resolution. However, actuating DNA devices with external fields has been limited to simple systems such as levers and hinges. This work seeks to enable direct manipulation of complex DNA nanodevice assemblies, in particular demonstrating actuation into many different states (i.e. assembly configurations), using DNA strand displacement, magnetic fields, and optical traps.

We employ scaffolded DNA origami (DO) for the fabrication of DNA assemblies. We designed these DO assemblies in the custom software MagicDNA, which enables the design of complex 3D assemblies on the order of minutes. Advanced 3D dynamic DO devices have expanded to larger scales through hierarchical assemblies that integrate many individual structures in a controlled way. Here, we present two reconfigurable higher order DO assemblies: 1) The 6-bar DO assembly demonstrates reconfiguration into many different conformations using DNA strand displacement and 2) the lattice DO assembly to demonstrate real-time assembly actuation with magnetic fields and optical tweezers.

The 6-bar DO assembly is a device comprised of 6-links with multiple strut regions that may lock the structure into pre-defined shapes. The struts may be displaced via DNA strand displacement and replaced to form a new shape. The free energies are explored to understand shape preference and used to guide the development of reconfiguration schemes. These shapes can polymerize into DO-tubes with stiff cross-sections using the same DNA strand displacement scheme as the individual devices. This work demonstrates higher order reconfiguration of a complex DO device is possible with DNA strand displacement understanding free energy of various states to guide tuning of structure and mechanical properties, which could enable tailoring of DNA assemblies for a variety of biological, chemical, or mechanical applications.

The latter portion of this work expands on the reconfiguration of higher order DO assemblies through external inputs. The primary lattice DO assembly is comprised of 4-links of a 3x4 cross-section of dsDNA helices with 16 ssDNA joint connections (4 per link) and a secondary lattice DO assembly comprised of 7-links with 6-links of a 3x4 cross-section of dsDNA helices and 1-link of a 2x6 cross-section with 32 ssDNA joint connections amongst the 7-links. Both lattice devices are designed with ssDNA overhangs in the corners of the joints that allow for actuation of the lattice, producing a more compact lattice structure. The actuation strands can also be decorated with fluorescent molecules and quenchers to provide a readout of the assembly configuration. Fluorometer results of the single-scaffold lattice show high efficiency of quenching of actuation strands resulting in successful actuation. Both lattice constructs are designed to polymerize into arrays when ssDNA polymerization strands are added. Our goal is to create micron-scale reconfigurable constructs that can be actuated via magnetic fields or optical trapping to expand the dynamic capabilities to DO to long-range micron-scale devices.

The reconfiguration and actuation of higher order DO assemblies provides tools for a variety of applications in the biological, chemical, and mechanical spaces. Creating bio-based tools for controlling more complex devices serving a foundation for nano-or micro-scale robotic systems provides platforms for multiplexing the control of nanomachines or molecular interactions. DNA strand displacement amongst large-scale DO assemblies shows local interactions among stiff DNA cross-sections can cause shape changes and external micron-bead actuation on DO lattice assemblies shows a level of spatiotemporal control for real-time manipulation of molecular systems. Investigating these actuation methods on higher order dynamic DO assemblies provides a basis for future actuation methods for nanoscale and micron-scale DNA constructs. Increased complexity of current actuation capabilities in the DNA nanotechnology space provides tools and methods spanning across multiple disciplines in engineering.

563d Kinetic and Parametric Studies of Pd Mineralization on Barley Stripe Mosaic Virus (BSMV) Virus-like-Particles (VLPs) As Biotemplates

Chamath Vindula Basnayake Pussepitiyalage, Shohreh Hemmati, Akash J. Vaidya, Che-Yu Chou, Kevin Solomon, Michael T. Harris, Sue Loesch-Fries

1-Dimensional (1D) metal nanostructures such as nanowires and nanorods are increasingly important in a wide array of applications ranging from electronics and catalysis to energy storage[1]. Sustainable green methods of nanostructure synthesis are gaining attention due to the negative environmental impact of conventional methods of producing nanostructures. Application of viral biotemplates such as Barley Stripe Mosaic Virus (BSMV) Virus-Like-Particles (VLPs) is one such promising green method for producing metal nanorods such as Pd nanorods[2]. BSMV VLP Capsid Proteins (CPs) contain amino acid residues consisting of electron rich functional groups such as hydroxyl, carboxyl, and amine groups capable of reducing Pd cation complexes present in an aqueous solution to induce Pd mineralization on the CPs[3]. Multiple cycles of mineralization of Pd on BSMV VLPs can be used to obtain uniform and smooth Pd nanorods. This study investigates the adsorption and reduction of Pd species on the BSMV VLP CPs using different palladium salt precursors (Na_2PdCl_4, Na_2PdBr_4, $PdCl_2$, and $Pd(NO_3)_2$) by the amino acid residues during each cycle of coating by the Ultraviolet-visible (UV-vis) Spectroscopy[3]. The effect of reaction conditions including palladium salt precursor and BSMV VLP concentration is examined using a parametric study. A kinetic study is used to calculate the reduction rate in each cycle of coating to propose a reaction mechanism for Pd mineralization on BSMV VLP. The observations from the kinetic and parametric studies are useful in optimizing Pd mineralization on the BSMV VLPs to obtain uniform Pd nanorods. The washing procedure for procuring Pd mineralized BSMV VLPs from the reaction solution used in each cycle of coating is a major challenge that has been addressed in this study.

This works is supported by the Nation Science Foundation (NSF) under grant number 2028634.

563e Breast Cancer Specificity Assessment of Tumor Targeted Nano-, Encapsulated Manganese Oxide (NEMO) Particles

Celia Martinez de la Torre, Dhruvi Panchal, Kasey Freshwater, Margaret Bennewitz

Gadolinium chelates are the gold standard contrast agents for breast magnetic resonance imaging (MRI), but their constant MRI signal and lack of specificity results in high background and misdiagnosis. This lack of targeting makes it challenging to differentiate between benign and malignant breast masses, resulting in high false-positive rates of up to 25%. We have developed an enhanced contrast agent that utilizes tumor-targeted pH-sensitive manganese oxide (MnO) nanoparticles, which in physiological pH, produce low signal ("OFF"). After cancer cell uptake and digestion within low pH endosomes/lysosomes, the MnO dissociates into Mn^{2+}, turning the MRI signal "ON."

Nano-, Encapsulated Manganese Oxide (NEMO) particles were synthesized by the encapsulation of nanocrystalline MnO within poly(lactic-co-glycolic acid) (PLGA) and poly(ethylene glycol) (PEG) via single emulsion. The nanocrystals and nanoparticles were chemically and physically characterized by TEM, XRD, FTIR, SEM, and DLS. After characterizing the particles, a fluorescent tumor-targeting peptide against underglycosylated mucin-1 (uMUC-1) was attached through click chemistry using copper (I) as the catalyst. MUC-1 was chosen as a tumor target, as it is overexpressed and underglycosylated on breast cancer cells, but not on benign or healthy cells. The uMUC-1 conjugated NEMO particles were characterized with DLS to assess hydrodynamic size (~190 nm) and fluorescence to ensure attachment of the targeting peptide. Before evaluating the cellular uptake of the NEMO particles, different control and breast cancer cell lines were characterized for expression of MUC-1 and uMUC-1 via flow cytometry. HEK293T (control), BT20 (triple negative), and T47D (luminal A) were selected for particle specificity studies, as their expression varied from no MUC-1 or uMUC-1 (HEK293T) up to 85% uMUC-1+ cells (T47D).

Cells were exposed to targeted NEMO particles (5 μg of Mn/mL) for 2 hours, followed by an MRI scan of the cell pellet and collected media. The change of R_1 (ΔR_1) between the control group (only exposed to media) and the nanoparticles-exposed group was calculated post imaging. In the case of cell pellets, HEK293T had a larger ΔR_1 than BT20 or T47D (266% vs. 116% vs. 162%). This was not expected based on relative MUC-1 and uMUC-1 expression. However, when scanning media alone, the opposite results were observed where HEK293T had a lower ΔR_1 than BT20 and T47D (13% vs. 67% vs. 56%), meaning that higher amounts of Mn^{2+} were found in BT20 and T47D media. Increased Mn^{2+} in the media could indicate that cancer cells internalized NEMO particles faster, dissolved MnO, and then released it. Therefore, further characterization of the time course of cellular uptake was performed by exposing the cells to NEMO particles (1.625 μg of Mn/mL) for 0.5, 1, and 2 hr, and again calculating ΔR_1. Lower expression of uMUC-1 resulted in a smaller change in MRI signal at shorter time points. Already at 30 min, T47D had over 200% ΔR_1 due to its high uMUC-1 expression and thus fast NEMO particle uptake, while it took 2hrs for the other two cell lines to reach that level. These results showed the dependency of uMUC-1 expression on NEMO particle uptake and, consequently, MRI signal.

Lastly, T47D tumor-bearing mice were injected with targeted NEMO particles (7.5 mg/kg), and MRI scans were performed pre-injection (baseline) and 1, 2, 4, and 24hr post-injection to assess the variation in MRI signal in the tumor. After 1hr and 2hr, the MRI signal intersity increased by ~37%

compared to baseline, followed by a slight decrease to ~18% above baseline at 24hr. Additionally, through abdominal MRI scans, signal intensity was visualized in the liver and kidneys after NEMO particle injection, highlighting possible elimination via these routes. Future biodistribution studies will be performed to correlate MRI signal to Mn organ content. Overall, targeted NEMO particles caused an increase in T_1 MRI signal in breast cancer cells expressing uMUC-1+ in vitro and in vivo, and further study is warranted to compare its sensitivity and specificity to Gd-chelates.

563f Award Submission: Next-Generation Tattoo-Ink for Improved Endoscopic Imaging

Subhadeep Dutta, Jordan Yaron, Rahul Pannala, Kaushal Rege, Mallikarjun Gosangi

Endoscopic tattooing (or commonly known as "colon tattooing") is a routinely used clinical technique to mark colorectal lesions (i.e. polyps) for successive surgical resection or to facilitate follow-up of the endoscopic resection site. Colon tattooing procedures are typically recommended for lesions with evident or suspected colorectal cancers; lesions with dimensions ≥2cm in diameter with potential risk of cancer, and for flat broad lesions removed by endoscopic submucosal dissection (ESD) or endoscopic mucosal resection (EMR) procedures. This technique involves injecting a dark-colored (i.e., high contrast) ink directly into the submucosal layer of the colon; or the ink can also be delivered through a saline bleb created in the submucosa. Currently used tattoo inks in clinic are prone to many side effects including peritonitis, abscess formation and / or development of inflammatory bowel disease. Additionally, submucosal fibrosis associated with existing commercial inks can interfere with endoscopic resection techniques later and may increase the risk for incomplete resection. These products can rapidly and substantially diffuse throughout the submucosal tissue and also to other tissues, which complicates the approximation of a tattoo with the intended lesion of interest. Hence, there is an unmet clinical need for new endoscopic tattoo inks which can (1) retain localization for improved follow-up accuracy and (2) do not induce toxic, inflammatory and / or fibrotic responses in long-term. To address this medical problem, we generated novel biomaterial-based tattoo-ink, formulated by combining (1) metal-derived nanoparticles (such as polysaccharide-coated negatively charged iron oxide nanoparticles) as the contrast agent and (2) mucoadhesive polymer biomaterials as a bioprotective coating and retention-mediating carrier, for applications in endoscopic imaging of the gastrointestinal tract.

Polysaccharide-coated iron oxide nanoparticles were synthesized using a one-pot co-precipitation method. A library of cationic biomaterial derivatives was prepared based on variable molecular weight, terminal functional group, and chemical functionalization including thiolation, quaternization. All biomaterial derivatives were characterized using FTIR spectroscopy; amine and thiol contents were measured using ninhydrin and elman's assay, respectively. A series of biomaterial-based tattoo-ink candidates were generated by modulating the ratio between the contrast agent and the biomaterial-based coating. Our tattoo-ink formulations, 0.25-2μm in hydrodynamic diameter, demonstrated near-equivalent contrast *ex vivo* with fresh porcine intestinal tissue as that shown by commercial inks, while also exhibiting substantially reduced diffusion in tissue, thus leading to better spot localization. The ability of these tattoo-ink constituents to induce IRF3 and NFκB inflammatory pathways in cells were evaluated *in vitro* using J774-DUAL mouse macrophage reporter cells and indicated minimal immunogenicity from these tattoo-ink constituents. Preclinical studies in immunocompetent Balb/c mice using subcutaneous injections indicate extensive diffusion of the commercial ink within the subcutis and significant leakage into the hypodermis. However, our formulations of biomaterial-based tattoo-ink demonstrated high contrast, effective spot localization in the subcutis and no leakage to other tissues, indicating better performance for visualization *in vivo*. Histopathology analyses on Handamp;E-stained sections revealed lead tattoo-ink candidates with biomaterials chemistry-dependent biocompatibility responses and very minimal immune response *in vivo*. Additionally, our tattoo-ink formulation was visualized under clinical MR-imaging system which enhances its potential use in multimodal imaging capability with non-invasive follow-up procedure. These biomaterial-based tattoo-ink formulations represent a next generation biocompatible endoscopic ink which will have measurable impacts on patient quality of life and procedural outcomes with a clear path to clinical translation.

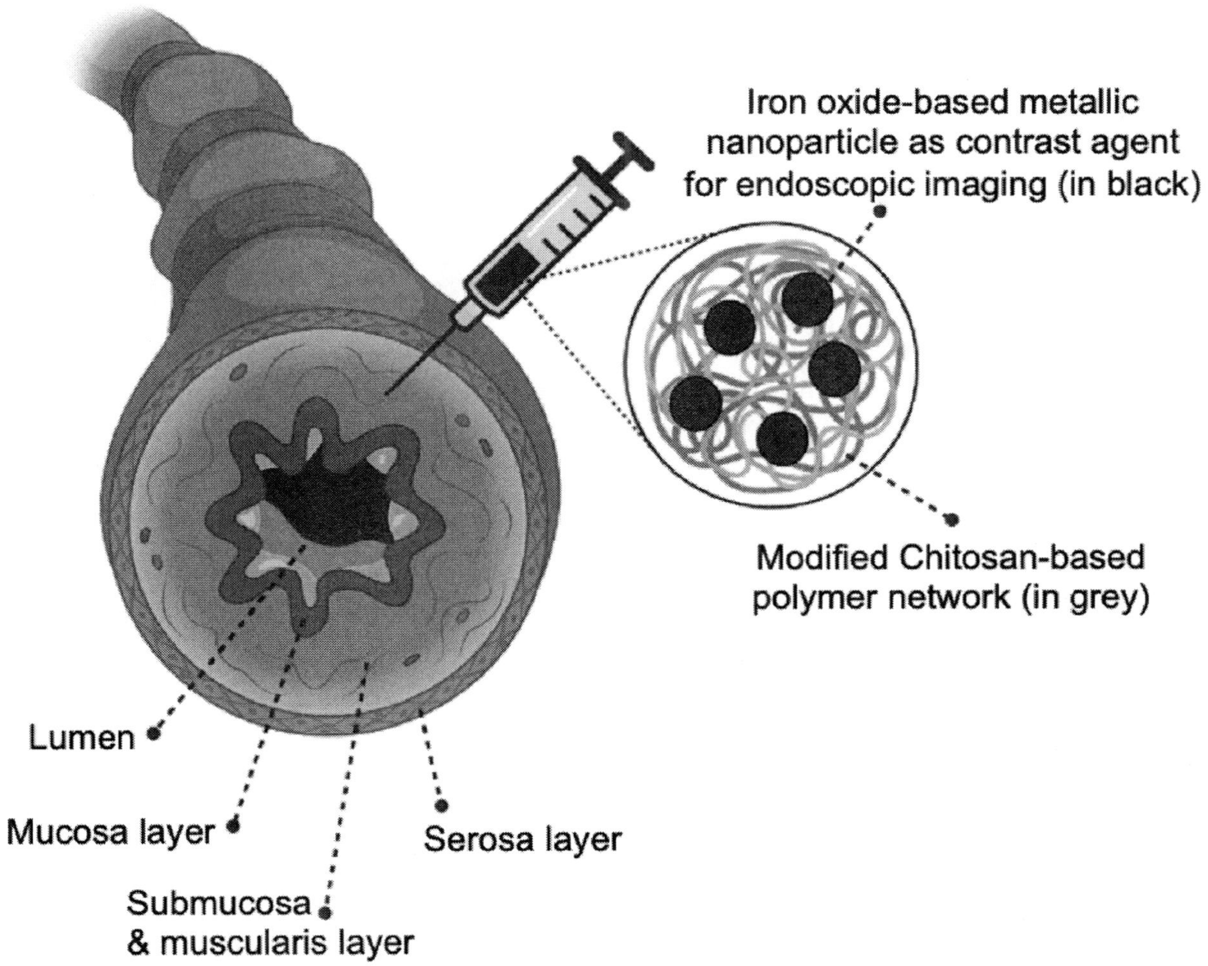
Iron oxide-based metallic
nanoparticle as contrast agent
for endoscopic imaging (in black)
Modified Chitosan-based
polymer network (in grey)
Lumen
Mucosa layer
Serosa layer
Submucosa
& muscularis layer

368a Surface-Mediated Assembly of Site-Modified Green Fluorescent Protein into Two-Dimensional Nanosheets As a Platform for Hierarchical Materials Fabrication

Nada Naser, Karthik S. Pushpavanam, Jinrong Ma, François Baneyx

Hierarchical architectures in which individual nanoscale components organize to determine order or function at larger scales have extraordinary potential for the creation of active interfaces and multidimensional structures with unique mechanical, optical, and catalytic properties. High information content macromolecules such as nucleic acids, foldamers, peptides and proteins, are desirable building blocks for such hierarchies due to their monodispersity and programmability. Among these species, proteins are especially valuable owing to their broad range of structures, chemistries and functions, and their amenability to genetic and chemical modifications. To date, the assembly of robust protein-based hierarchies has been challenging. Here, we use a variant of superfolder Green Fluorescent Protein that is modified with 4 equatorial cysteines (sfGFP(4C)) and functionalized with a thiol-reactive linker to mediate efficient 2D assembly. Further, surface interactions are mediated by a solid-binding sequence that is genetically inserted within a permissive loop of the protein. In addition to its inherent optical properties, this system is being used as a platform for the bottom-up construction of active interfaces via oriented integration of light-activated proton pumps solubilized in peptidiscs.

368b A Machine Learning Study on the Result of Polarizable Molecular Dynamics of Ionic Liquid-Based Solid Polymer Electrolytes for Li^+-Ion Batteries By Graph Dynamical Networks

Chanui Park, Anseong Park, Sebin Kim, Seunghyok Rho, Minhwan Lee, Sangwoo Kwon, JunBeom Cho, Seulwoo Kim, Won Bo Lee

Solid polymer electrolytes (SPEs) are promising candidates for safe and stable lithium-ion secondary battery electrolytes. Poly(ethylene oxide) (PEO) has been widely used as a polymer host, and room temperature ionic liquids (RTILs) have been added to significantly increase lithium ion mobility in SPEs. Our study focused on exploring the role of each component and geometrical condition in the electrolytes from simple ionic liquid electrolytes to multicomponent complex SPEs: LiTFSI/PYR_{14}TFSI binary bulk, LiTFSI/PYR_{14}TFSI/PEO ternary bulk, and LiTFSI/PYR_{14}TFSI mixture in a PEO-branched cylindrical nanopore. Molecular dynamics (MD) simulations of these systems were performed previously [1] under the APPLEandamp;P polarizable force field.

An unsupervised learning method called Graph Dynamical Networks (GDyNets) [2] was suggested for learning atomic scale dynamics from MD trajectories of arbitrary condensed phases. GDyNets can classify local configurational states around a target atom using graph convolutional neural networks and obtain transition probabilities between the states based on the Koopman model [3].

In this study, the MD trajectory of each electrolyte was trained with GDyNets where lithium ions were determined as target atoms. For each trajectory, we measured several state-wise and transition-wise properties of all Li^+ ions from the trained model. Population of each classified state was calculated. State-weighted radial and spatial distribution functions around the Li^+ ions identified the local structure of each state. The Koopman matrix was evaluated after the lag time was determined. From the eigenvector decomposition of the Koopman matrix, the slowest transitions were identified, and their timescales were evaluated. The Chapman-Kolmogorov test was performed to check whether the trained model is Markovian. Conductivity of each transition revealed that which transitions have major contributions to the overall conductivity. In addition to properties of all Li^+ ions, we analyzed individual ion trajectories with the same model, and ions with similar behavior were clustered into a group. Our study contributes to elucidate lithium-ion acceleration conditions in SPEs and give a systematic method to analyze multicomponent MD trajectories with machine learning techniques.

References

[1] Kim, Seulwoo, et al. 'Molecular dynamics study on lithium-ion transport in PEO branched nanopores with PYR_{14}TFSI ionic liquid.' *Battery Energy* (2022): 20210013.

[2] Xie, Tian, et al. 'Graph dynamical networks for unsupervised learning of atomic scale dynamics in materials.' *Nature Communications* 10.1 (2019): 1-9.

[3] Koopman, Bernard O. 'Hamiltonian systems and transformation in Hilbert space.' *Proceedings of the National Academy of Sciences of the United States of America* 17.5 (1931): 315.

368c Lithium Ion Transport Mechanism in PYR_{14}tfsi/PEO Branched Nanopore System: A Polarizable Molecular Dynamics Study

Sebin Kim, Seulwoo Kim, Minhwan Lee, Chanui Park, Anseong Park, Sangwoo Kwon, JunBeom Cho, Seunghyok Rho, Won Bo Lee

A polarizable molecular dynamics (MD) simulation study was conducted to verify the lithium ion transport mechanism in polymer branched ion channels filled with ionic liquids. Specific compounds of the electrolytes comprised the solution of [lithium][bis(trifluoromethanesulfonyl)imide] (LiTFSI) in [1-butyl-1-methylpyrrolidinium][bis(trifluoromethanesulfonyl)imide] (PYR_{14}TFSI) and poly-ethylene oxide (PEO) were branched at the wall of nanopores at regular intervals. MD simulations with polarizable force fields, especially APPLEandamp;P force fields [1], have been performed to precisely analyze Li ion transport mechanism. The effects of the ratio of Li ions to PEO chains and size of nanopores on the structural and dynamic properties of Li ions were examined.

As a result of structural analysis, Li ions prefer to migrate in the longitudinal direction due to the confinement in the radial direction. Interestingly, it was found by cluster analysis that the complex of Li and TFSI ions preferred rather than in the PEO system, although is the most prominent structure in the binary electrolytes (LiTFSI/PYR_{14}TFSI) of similar salt concentration. [2,3] With smaller Li-TFSI complex, the mobility of Li ions was expected to be higher than binary electrolyte systems, and these results were consistent with the result of dynamic properties; Diffusion coefficients and ionic conductivities of Li ions in the longitudinal direction was improved in the PEO system. Particularly, as the ratio of the Li ions to PEO chains increases and the pore size decreases, the mobility of Li ions increases. We also investigated the residual time of Li ions in the complex to analyze transport mechanism of Li ions. In the PEO system, residual time of Li ions for TFSI was decreased compared to the binary electrolytes system. In addition, as a result of single ion trajectory analysis, Li ions encountered more TFSI ions during simulation times in the PEO system.

Ref.

[1] Borodin O., Polarizable force field development and molecular dynamics simulations of ionic liquids. *J. Phys. Chem. B* 113, 11463–11478 (2009).

[2] Tong J, Wu S, von Solms N, et al. The effect of concentration of lithium salt on the structural and transport properties of ionic liquid-based electrolytes. Front Chem. 2020;7:1-10.

[3] Haskins JB, Bennett WR, Wu JJ, et al. Computational and experimental investigation of Li-doped ionic liquid electrolytes: [pyr14][TFSI], [pyr13][FSI], and [EMIM][BF4]. J Phys Chem B. 2014;118:11295-11309.

368d Improvement of Storage Stability of mRNA Vaccine Using Lipid Based Drug Delivery System

MinJeong Kim, Taekyoung Lee, Hyunjin Kim, Sangmin Lee, Minsub Chung

The lipid nanoparticle (LNP) carrier-based mRNA vaccine was first used as the COVID-19 vaccine, proving to be a next-generation vaccine technology beyond the existing vaccine. However, as it is the first mRNA vaccine, problems were also exposed, especially low stability of the mRNA and lipid nanoparticle carriers and need to be stored/transmitted at extremely low temperatures, such as cold-chain issues and mRNA distribution problems in underdeveloped countries. Therefore, the purpose of this study is to develop an efficient and reliable storage system that can solve transportation and storage stability, which has been pointed out as a problem of existing LNPs for mRNA delivery.

In this study, through the low-temperature decompression concentration method and stabilizer, our goal is maintaining more than 90% of the initial efficacy of the mRNA vaccines at low temperature (2 ~ 8 °C) for more than 3 months. The developed low-temperature decompression concentration method and stabilizing agents enabled made mRNA vaccines more stable in low temperature (2 ~ 8 °C). Comparative study to existing LNP system and conventional storage protocol was presented to monitor the change of LNP size, distribution, and efficacy data over time. Preservation additives and freeze-drying process was optimized to improve the storage period under refrigerated distribution conditions. mRNA vaccines in good stability will reduce the difficulties caused by the existing transportation and storage problems and reduce the disposal of the vaccines, contributing to the development of vaccines against spread of infectious diseases or cancer cells.

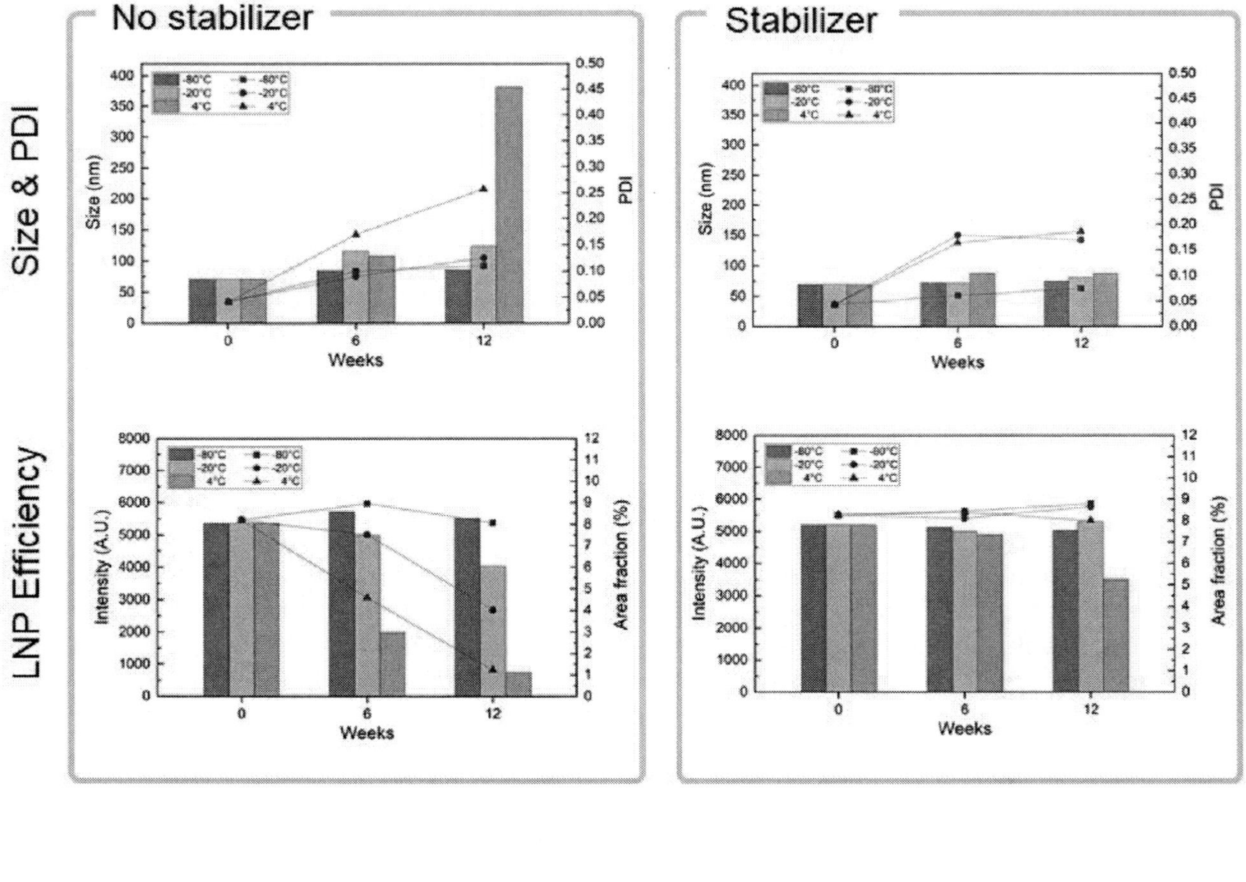

368e Synthesis of Metal-Organic Frameworks (MOFs) and Evaluation of Their Toxicological Profiles.

Olivia Rose, Yon Rojanasakul, Cerasela Zoica Dinu

Metal-organic frameworks (MOFs), materials under the nanotechnology umbrella, are hybrids formed through the combination of metal ions and organic linkers[1,2]. The integration of these two constituents allows for tunable porosity, biocompatibility, adaptable shape, high thermal stability and surface area[1,3,4] to be introduced at their synthesis to subsequently result in controlled structure and physicochemical characteristics that make them attractive to be employed in a variety of applications from gas storage[5,6,7], to catalysis[8,9,10], and drug delivery[11,12,13,14] as well as contrast agents for bioimaging[15,16,17].

The exposure and implementation of MOFs however needs to be evaluated because of possible antagonistic interactions of such frameworks with given biosystems. For instance, **Su** et al. utilized *in vitro* toxicity evaluations to assess the cytotoxic effects of medi-MOF-1. Pancreatic cancer cells (BxPC-3) were exposed to the MOFs and assessed using the 3-(4,5-dimethylthiazol-2-yl)-2,5-diphenyltetrazolium bromide (MTT) assay. Results showed that there was a cell growth inhibition in a dose-dependent manner, specifically, exposure to 50 μg/ml MOFs caused the highest levels of cell death[18]. **Chen** et al., exposed MIL-160(Fe) MOFs to human normal liver and hepatocellular carcinoma cells. A decrease in cell viability in dose-dependent manner was observed and the cause of toxicity was due to the MOFs' metal component[19]. While these *in vitro* toxicity studies helped correlate toxicological pathways function of both MOFs' physicochemical characteristics and their synthesis method cyto- and genotoxic effects and their possible synergism in how they account for homeostatic cellular changes have been sparse. Moreover, the ability to identify any cell generational transformation and propagation of such transformation upon framework exposure is currently lacking. Further, *in vivo* toxicity evaluations were focused on limited MOFs and only revealed weak mechanistic analysis or minimal toxicological pathways identification. Without such correlations and analysis, framework deleterious effects (when individual or when in synthesized structures) are difficult to assess and control. As such, lack of analysis and evaluation of toxicological profiles could jeopardize the implementation of the frameworks in consumer applications.

This project aims to investigate the toxicological profiles of selected MOFs and their thin films to determine mechanisms of toxicity before consumer implementation. The objective is to contribute to the understanding of MOFs deleterious effects, both using cyto- and genotoxic approaches as well as real-time, high throughput strategies. The proposed methods involve synthesizing and characterizing selected MOFs and their thin films and exposing such frameworks and films to human epithelial cells, where the lung cells serve as model for the toxicity of target system (i.e., lungs). Prior to exposure, the MOFs/thin films are characterized by microscopical and spectroscopical techniques to determine their physicochemical characteristics. Subsequently, upon exposure frameworks, films and cell systems are investigated using both microscopy and cellular assay. Our results show that the frameworks affect cells in a dose-dependent manner with such effects being function of the materials' physico-chemical properties. Possible mechanisms responsible for cell toxicity are also derived from systems characteristics. Such studies could not only lead to rationale, safe-by-design strategies development of nanomaterials and nanomaterials films for biotechnological implementation but further, could also allow for user-controlled implementation of such structures.

References:

(1) Lin, W. X.; Hu, Q.; Yu, J. C.; Jiang, K.; Yang, Y. Y.; Xiang, S. C.; Cui, Y. J.; Yang, Y.; Wang, Z. Y.; Qian, G. D. Low Cytotoxic Metal-Organic Frameworks as Temperature-Responsive Drug Carriers. *Chempluschem* **2016**, *81* (8), 804-810, Article. DOI: 10.1002/cplu.201600142.

(2) Tamames-Tabar, C.; Cunha, D.; Imbuluzqueta, E.; Ragon, F.; Serre, C.; Blanco-Prieto, M. J.; Horcajada, P. Cytotoxicity of nanoscaled metal-organic frameworks. *Journal of Materials Chemistry B* **2014**, *2* (3), 262-271, Article. DOI: 10.1039/c3tb20832j.

(3) Yang, B. C.; Shen, M.; Liu, J. Q.; Ren, F. Post-Synthetic Modification Nanoscale Metal-Organic Frameworks for Targeted Drug Delivery in Cancer Cells. *Pharmaceutical Research* **2017**, *34* (11), 2440-2450, Article. DOI: 10.1007/s11095-017-2253-9.

(4) Lucena, M. A. M.; Oliveira, M. F. L.; Arouca, A. M.; Talhavini, M.; Ferreira, E. A.; Alves, S.; Veiga-Souza, F. H.; Weber, I. T. Application of the Metal-Organic Framework Eu(BTC) as a Luminescent Marker for Gunshot Residues: A Synthesis, Characterization, and Toxicity Study. *Acs Applied Materials andamp; Interfaces* **2017**, *9* (5), 4684-4691, Article. DOI: 10.1021/acsami.6b13474.

(5) Kayal, S.; Sun, B. C.; Chakraborty, A. Study of metal-organic framework MIL-101(Cr) for natural gas (methane) storage and compare with other MOFs (metal-organic frameworks). *Energy* **2015**, *91*, 772-781, Article. DOI: 10.1016/j.energy.2015.08.096.

(6) Gutov, O. V.; Bury, W.; Gomez-Gualdron, D. A.; Krungleviciute, V.; Fairen-Jimenez, D.; Mondloch, J. E.; Sarjeant, A. A.; Al-Juaid, S. S.; Snurr, R. Q.; Hupp, J. T.; et al. Water-Stable Zirconium-Based Metal-Organic Framework Material with High-Surface Area and Gas-Storage Capacities. *Chemistry-a European Journal* **2014**, *20* (39), 12389-12393, Article. DOI: 10.1002/chem.201402895.

(7) Bunzen, H.; Kalytta-Mewes, A.; van Wullen, L.; Volkmer, D. Long-term entrapment and temperature-controlled-release of SF6 gas in metal-organic frameworks (MOFs). *Beilstein Journal of Nanotechnology* **2019**, *10*, 1851-1859, Article. DOI: 10.3762/bjnano.10.180.

(8) Ghorbanloo, M.; Safarifard, V.; Morsali, A. Heterogeneous catalysis with a coordination modulation synthesized MOF: morphology-dependent catalytic activity. *New Journal of Chemistry* **2017**, *41* (10), 3957-3965, Article. DOI: 10.1039/c6nj04065a.

(9) Valvekens, P.; Jonckheere, D.; De Baerdemaeker, T.; Kubarev, A. V.; Vandichel, M.; Hemelsoet, K.; Waroquier, M.; Van Speybroeck, V.; Smolders, E.; Depla, D.; et al. Base catalytic activity of alkaline earth MOFs: a (micro)spectroscopic study of active site formation by the controlled transformation of structural anions. *Chemical Science* **2014**, *5* (11), 4517-4524, Article. DOI: 10.1039/c4sc01731e.

(10) Ling, P. H.; Lei, J. P.; Zhang, L.; Ju, H. X. Porphyrin-Encapsulated Metal-Organic Frameworks as Mimetic Catalysts for Electrochemical DNA Sensing via Allosteric Switch of Hairpin DNA. *Analytical Chemistry* **2015**, *87* (7), 3957-3963, Article. DOI: 10.1021/acs.analchem.5b00001.

(11) Kundu, T.; Mitra, S.; Patra, P.; Goswami, A.; Diaz, D. D.; Banerjee, R. Mechanical Downsizing of a Gadolinium(III)-based Metal-Organic Framework for Anticancer Drug Delivery. *Chemistry-a European Journal* **2014**, *20* (33), 10514-10518, Article. DOI: 10.1002/chem.201402244.

(12) Pan, Y. B.; Wang, S. Q.; He, X. C.; Tang, W. W.; Wang, J. H.; Shao, A. W.; Zhang, J. M. A

combination of glioma in vivo imaging and in vivo drug delivery by metal-organic framework based composite nanoparticles. *Journal of Materials Chemistry B* **2019**, *7* (48), 7683-7689, Article. DOI: 10.1039/c9tb01651a.

(13) Orellana-Tavra, C.; Marshall, R. J.; Baxter, E. F.; Lazaro, I. A.; Tao, A.; Cheetham, A. K.; Forgan, R. S.; Fairen-Jimenez, D. Drug delivery and controlled release from biocompatible metal-organic frameworks using mechanical amorphizationt. *Journal of Materials Chemistry B* **2016**, *4* (47), 7697-7707, Article. DOI: 10.1039/c6tb02025a.

(14) Shu, F. P.; Lv, D. J.; Song, X. L.; Huang, B.; Wang, C.; Yu, Y. Z.; Zhao, S. C. Fabrication of a hyaluronic acid conjugated metal organic framework for targeted drug delivery and magnetic resonance imaging. *Rsc Advances* **2018**, *8* (12), 6581-6589, Article. DOI: 10.1039/c7ra12969f.

(15) Taylor, K. M. L.; Rieter, W. J.; Lin, W. B. Manganese-Based Nanoscale Metal-Organic Frameworks for Magnetic Resonance Imaging. *Journal of the American Chemical Society* **2008**, *130* (44), 14358-+, Article. DOI: 10.1021/ja803777x.

(16) Taylor-Pashow, K. M. L.; Della Rocca, J.; Xie, Z. G.; Tran, S.; Lin, W. B. Postsynthetic Modifications of Iron-Carboxylate Nanoscale Metal-Organic Frameworks for Imaging and Drug Delivery. *Journal of the American Chemical Society* **2009**, *131* (40), 14261-+, Article. DOI: 10.1021/ja906198y.

(17) Chowdhuri, A. R.; Bhattacharya, D.; Sahu, S. K. Magnetic nanoscale metal organic frameworks for potential targeted anticancer drug delivery, imaging and as an MRI contrast agent. *Dalton Transactions* **2016**, *45* (7), 2963-2973, Article. DOI: 10.1039/c5dt03736k.

(18) Su, H. M.; Sun, F. X.; Jia, J. T.; He, H. M.; Wang, A. F.; Zhu, G. S. A highly porous medical metal-organic framework constructed from bioactive curcumin. *Chemical Communications* **2015**, *51* (26), 5774-5777, Article. DOI: 10.1039/c4cc10159f.

(19) Chen, G. S.; Leng, X.; Luo, J. Y.; You, L. T.; Qu, C. H.; Dong, X. V.; Huang, H. L.; Yin, X. B.; Ni, J. In Vitro Toxicity Study of a Porous Iron(III) Metal-Organic Framework. *Molecules* **2019**, *24* (7), 14, Article. DOI: 10.3390/molecules24071211.

368f Diffusion Growth Mechanism of Penta-Twinned Ag/Cu Nanowires: Multiscale Theory

Jianming Cui

Penta-twinned metal nanowires are finding applications in many different technologies, ranging from health monitoring and human-machine interfaces to smart fabrics to windows that change color or tint in response to light, provide shielding from electromagnetic interference, and something about heating. These nanowires find applications in catalysis, photothermal desalination, transparent and flexible electronics applications, such as touch screens, and various elements of solar cells.

The penta-twinned nanowires can be produced with high aspect ratios from diverse face-centered cubic (fcc) metals, with {100} as sides and {111} as ends. However, the mechanism of their anisotropic growth is still poorly understood. We develop a model combining deposition and surface diffusion to predict the final aspect ratio of Ag and Cu nanowires. We describe two aspects of nanowire growth, denoted as the seed stage and the wire stage. In the seed stage, the deposition is much slower than the surface diffusion, so that the seed shape only depends on the difference between interfacet diffusion rates. In the wire stage, the nanowires continue to grow longer until the diffusion is slower than deposition.

The structure of nanowire seeds is similar to Marks decahedra. In the uncapped Ag case, the diffusion from {100} to {111} is comparable to the opposite and both inter-facet diffusion processes are faster than deposition in seed stage. The result shows that the uncapped Ag seeds could lead to three kinds of products with different strucutures as nanowires, decahedron particles and giant decahedron seeds. For the nanowire products, our predictions well match the experimental results. In the uncapped Cu case, we found out that the diffusion from {100} to {111} is slower than the opposite, which impies that the seeds would prefer to grow to large decahedron particles. In experiments, the Cu nanowires were always synthesized with some surfactants. In the case of Cu seeds with Cl on the surface, the relationship between the inter-facet diffusion processes are inversed. The diffusion from {100} to {111} is the fastest and the diffussion from {111} to {100} is even slower than the deposition, which means that the Cu seeds with Cl would prefer to grow to wire products. Based on our model, the predictions are able to represent the average sizes from experimental results.

368g Protein Templated Core/Shell Nanostructures for Photothermal Therapy and SERS Mediated Intracellular ROS Detection

Animesh Pan, Muzahidul Islam Anik, Md Golam Jakaria, Samantha Meenach, Geoffrey D. Bothun

Core/shell gold 'raspberry' nanostructures capable of multiple therapeutic functionalities were synthesized using a template comprised of monodispersed anionic protein (bovine serum albumin) nanoparticles coated with a cationic biopolymer (poly-L-lysine). The nanostructures exhibited high photothermal conversion efficiency when exposed to near-infrared (NIR) laser which led to significant cellular inhibition of A549 human lung cancer cells due to intracellular hyperthermia. The raspberry structures also provided hotspots for surface-enhanced Raman scattering (SERS) ratiometric sensing of intracellular reactive oxygen species (ROS) when modified with the Raman reporter molecule 4-aminothiophenol (4-ATP). ROS was detected in A549 lung cancer cells upon photothermal heating of internalized nanostructures, enabling a possible mechanism for feedback on therapeutic efficacy. This was confirmed by adding the antioxidant N-acetylcysteine (NAC) and using a complimentary fluorescence technique, which showed that the amount of detectable intracellular ROS decreased. These safe-by-design gold raspberry nanostructures could be promising for simultaneous therapeutic applications and monitoring therapeutic efficacy.

368h Optimizing Composition and Solar Light Conditions for the Reversible Diels-Alder Reaction in Titanium Nitride Nanoparticle-Laden Epoxy

Madeline Finale, Kavon Mojtabai, Arnob Dipta Saha, Sanchari Chowdhury

Thermoset polymers such as epoxy resins have been used extensively in coatings, structural, electronics, and adhesive applications. Unlike thermoplastic polymers, the end-of-life products of these polymers cannot be repaired or recycled simply by applying heat as this will cause the material to be damaged rather than melt. We synthesized a reversible epoxy containing Diels-Alder adducts that is capable of depolymerizing at higher temperatures and repolymerizing at lower temperatures, allowing the epoxy to become recyclable. Additionally, photothermal plasmonic titanium nitride nanoparticles are incorporated to the epoxy matrix. When irradiated with solar light, the nanoparticles will induce heating at the nanoscale that is rapidly delivered to the Diels-Alder adducts, causing quick depolymerization in a specific area if desired. These epoxy composites were irradiated with different levels of light intensity from a solar simulator; light intensity was varied at different levels ranging from 2.445 mW/cm^2 to over 500 mW/cm^2. Reversible epoxies/nanoparticle composites containing different wt.% of TiN nanoparticles (0, 0.2, 0.5, and 0.7 wt.%) were tested for their light absorbance as well as photothermal efficiency. Optical microscopy is used to study the in situ melting of the epoxy composites under solar simulator. Both the nanoparticle loading as well as the light intensity were optimized to determine the optimal conditions for melting, which is important for the reprocessing and recycling of the epoxy. Results indicate that solar light concentrated with a lens can produce sufficient intensity to melt TiN nanoparticle-laden epoxy. In some cases, the sample showed signs of melting after 10 seconds of exposure to light, and at a temperature around 90°C. Such findings suggest that this procedure could be increased from the laboratory scale and used to extend the lifetime or recycle products.

368i Synthesis of Near-Infrared Pigments for Novel Sensor Applications

John Clark, Holly A. Stretz, Agoston Kiss

Calcium copper silicate, known as Egyptian Blue (EB), is a pigment that exhibits a strong near-IR fluorescent emission when exposed to a strong light source. This property makes it a novel candidate for use as a basis for dissolved ion sensors, as its IR signal is very strong in comparison with current commonly used IR reporters, such as gold shell nanoparticles. Most commercially available samples; however, are created with dyeing and pigmentation in mind, rather than chemical application. To apply EB as a sensor lens component, it must be synthesized to form a reproducible and laboratory grade sample, free from compounds present in dyes that can affect the adherence and emission of an EB coating on a lens. In order to synthesize a sample of EB, a solid-state reaction was employed to react powdered crystalline solids under a high temperature environment.

369a Upgrading Food Waste to High Commercial Value Chemicals

Yagya Gupta, Laura Elizabeth Beckett, Sunitha Sadula, Vibin Varghese, LaShanda Korley, Dionisios Vlachos

The profound challenge of food waste (FW) is evidenced by the fact that approximately a third of global food production (*i.e.,* 1.3 billion tons) ends up as waste. Traditional FW treatment methods, such as disposal into landfills and composting, are environmentally unfriendly. Intensive exploration of other methods that selectively convert the diverse components of FW to high commercial value chemicals is a key to alleviating this challenge. Extraction of antioxidant and antimicrobial activity providing chemicals (flavonoids, phenolic acids, etc.) from FW feedstocks is one such economically lucrative option to valorize FW. However, separation and purification of these high commercial value chemicals is challenging owing to their similar chemical nature, high boiling points and low concentrations in FW. To the best of our knowledge, currently there is no method to separate and purify phenolic compounds post extraction from FW or non-food biomass. Our study directly addresses this issue by proposing a scalable and extensible methodology for separation of the target molecules using molecular imprinted polymers. Synthesis of polymers with imprinted functionality is key in optimizing the selectivity and performance of the material. Computational methods were used to screen through 28 monomers to expose their interaction with the target molecule. Experimental investigations revealed that itaconic acid with chlorogenic acid as the template provides the highest separation factor (higher than the literature standard, acrylamide). The performance of the polymer at separating structurally similar was tested in various extraction solvents. Complementing IR studies were done to understand the underlying interactions. The application of the polymer for separation of polyphenols extracted from real food waste and standard mixtures with known composition is demonstrated.

369b Controlling Metal Nanoparticle Size Distribution through Microreactor Residence Time Distribution

Faiz Khan, Jessica Winter

The application of microreactors for nanoparticle synthesis has been widely studied and commercially implemented for applications in vaccine manufacturing and drug delivery. Because of their small length scale, microreactors systems offer the potential to achieve rapid convective mass transport in a fraction of seconds. This allows control of the nucleation and growth kinetics of nanoparticles. Besides mixing, nanoparticle size distribution is also affected by the residence time distribution of the reactor [1]. Conventional microreactor designs, including T-mixer and jet mixing reactor designs typically yield a laminar flow profile. Radial mixing under laminar conditions is limited to diffusion, leading to a broad residence time distribution (RTD) of particles inside the reactor. Longer residence time inside the reactor allows existing nanoparticles to grow, widening the particle size distribution. A polydisperse nanoparticle distribution compromises the uniformity of size-dependent nanoparticle properties such as fluorescence, catalytic activity, and surface plasmon resonance. A clear understanding of the RTD effect on product functionality is lacking. Hence, this research aims to investigate the effect of microreactor design on the RTD and a model nanoparticle system.

Our strategy involves producing nanoparticles using a jet mixing reactor (JMR) microreactor. The JMR is a three inlet microreactor, with two cross-flow jet streams impinging on a central main fluid stream. Two variants of JMR outlet designs were probed: (a) a curved helical outlet, and (b) a straight outlet. JMR RTDs were probed using a step input of rhodamine B tracer dye. The outlet solution absorbance was monitored at 565 nm using flow cells to compute the mean, variance, and skew of the RTD curve. The Villermaux-Dushman competitive reaction set was used to investigate the effect of mixing time for straight vs. curved outlets. Further insights into JMR mixing and RTD were found through COMSOL simulations. Gold nanoparticles were selected as the model reaction system due to their size-dependent surface plasmon resonance properties. The effect of RTD and inlet flow rates on nanoparticle size distribution was analyzed using transmission electron microscopy. Polydispersity index served as the evaluation criteria for JMR synthesis with curved versus straight outlets.

References:

[1] Gao, Yunhu, Bruno Pinho, and Laura Torrente-Murciano. 'Tailoring the size of silver nanoparticles by controlling mixing in microreactors.' Chemical Engineering Journal 432 (2022): 134112.

369c Development and Characterization of Recyclable Epoxy/Refractory Plasmonic Nanoparticles for Additive Manufacturing

Arnob Dipta Saha, Kavon Mojtabai, Madeline Finale, Samantha Lindholm, Brandon McReynolds, Youngmin Lee, John McCoy, Sanchari Chowdhury

Thermoset epoxy polymers are non-recyclable due to their irreversible crosslinked structure. We develop thermally reversible epoxies by incorporating dynamic covalent bonds like diene-dienophile linkages initiating Diels' Alder reaction chemistry. However, the lack of localized heating precision and low thermal conductivity of epoxies can make the recycling using bulk heating process ineffective. We apply refractory plasmonic inexpensive Titanium Nitride (TiN) for converting the incident light to localized heat leading to depolymerization of epoxies at targeted areas. Titanium nitride nanoparticles show strong plasmonic properties due to their metallic behavior and is attractive due to their superior thermal stability and broader solar light absorption features. The incorporation of titanium nitride nanoparticles in the reversible epoxies provide several advantages: (1) application of solar light to liquefy(depolymerize) the epoxy for reprocessing, (2) targeted repair of defects on a 3-D printed epoxy structure, (3) fast curing of epoxy using light, required for their 3-D printing applications, and finally (4) smoothening the rough surface of the final products. We are functionalizing TiN nanoparticles with thiol group via sonochemical method, for better dispersion of TiN nanoparticles in the epoxy matrix. The functionalization of nanoparticles is confirmed using FTIR spectroscopy. We also characterized the nanoparticles dispersion in the composites and the surface roughness of the composite structures using Atomic Force Microscopy. The surface roughness of the structure could be reduced by 45% by melting the top surface of the structure using solar irradiation.

369d Development of Magnetic Nanoparticles and Nanocomposites for Environmental and Biomedical Applications

Pranto Paul, J. Zach Hilt

Magnetic nanoparticles (MNPs) and their nanocomposites can provide unique solutions to environmental and biomedical problems. The magnetic nanoparticles enable for energy to be delivered remotely to the material using an alternating magnetic field (AMF). This localized energy can be used to drive processes to enhance material performance, including disruption of binding interactions, increased reactivity, manipulation of biological systems and assemblies, etc. Here, we will highlight a few diverse applications (e.g., low energy sorbent regeneration, enhanced reactivity, virus inactivation, etc.). For example, decades of use of halogenated chemicals in product manufacturing and commercial applications have been connected to detrimental health impacts. In response to these, we have applied magnetic nanocomposite for the reversible capture of halogenated contaminants such as perfluoroalkyl substances (PFAS) by regenerating the nanocomposite through an AMF.

369f Development of Methods for Precise, Multifactor Tuning of Shell Morphology on Silica-Encapsulated Gold Core-Shell Nanoparticles

Ellis Hammond-Pereira, Zengran Sun, Steven Saunders

A seeded encapsulation synthesis method was developed to facilitate precise, multifactor tuning of mesoporous silica-encapsulated gold core-shell nanoparticles (Au@SiO_2 CSNPs). The new seeded method prevents the Stöber synthesis portion of the reaction from impacting the gold core diameter. Altering the Stöber synthesis conditions thusly results in core-shell nanoparticles with different shell morphologies, without affecting the gold active phase. Because of this, remarkably simple catalytic studies can be performed to investigate the specific impact of pore morphology on catalytic performance. Both silica pore diameter and shell thickness (corresponding to pore length) were varied. It was found increasing ethanol concentration resulted in higher pore diameters, while reducing total solvent volume resulted in high shell thicknesses.

Rotational diffusion NMR studies using d-toluene revealed anisotropic reduction in rotational diffusion within the silica mesopores: rotation about the major molecular axis was reduced less than orthogonal rotation, indicating species tended to orient themselves beneficially within the pores. It was found the difference in rotation between axes was lessened as pore diameter increased. When catalyzing solvent-free benzyl alcohol oxidation, CSNPs with narrower pores demonstrated higher catalytic activity. CSNPs with longer pores also demonstrated higher activity, suggesting longer diffusion paths can increase the extent of beneficial orientation. Results from developing this method not only demonstrate clearly the value of mesoporous encapsulation, but also provide a simple, tunable platform for investigating separately the impacts of diffusion and reaction on catalysis.

369g NSEF Poster Session: Computational Studies on the Structural Properties of Square Colloids with Offset Magnetic Dipoles

Matthew Dorsey, Orlin D. Velev, Carol Hall

In recent work, Velev and coworkers have developed a new class of engineered materials that interact, assemble, reconfigure, and propel in response to external magnetic and electric fields. Cubic microparticles with a ferromagnetic-metallic coating on one or two opposing faces retain residual polarization when exposed to an external magnetic field, even after the field is turned off. The many different interactions that exist when anisotropic, magnetically-polarized colloidal particles are placed in tunable, external fields creates numerous design variables that are challenging to fully explore experimentally. The search for potentially useful structures formed by this colloidal material can be enhanced by computer simulations of colloidal assembly. In this work, we used Discontinuous Molecular Dynamics (DMD) to simulate the behavior of large systems of dipolar squares in the absence of a magnetic field. DMD is a fast variant of standard molecular dynamics that is applicable to systems of molecules interacting via discontinuous potentials, and is best suited for exploring phenomena that occur at long time scales. Microcubes were represented in quasi-2D as groupings of hard discs bonded together to create a rigid square geometry. Magnetic dipoles were mimicked in silico by embedding opposite electrostatic charges along one cubic face. Annealing, or "slow-cooling", simulations were performed in the absence of a magnetic field using the model described above to discover the equilibrium conformations of the dipolar squares.

We find that, as the strength of the dipolar interactions between squares overcomes the system's thermal energy, the dipolar squares assemble into single- or double-stranded assemblies, each with unique structures and phase diagrams in the temperature-density plane. Single-stranded assemblies of dipolar squares are associated with the formation of a percolated, or gel-like, state, while double stranded assemblies are associated with the formation of a nematic state. Using pair-wise potential energy calculations between dipolar squares, we show that whether a system of dipolar squares assembles into single- or double-stranded conformations depends on how the dipole is embedded within the square. Furthermore, by parameterizing the location of the dipole within the square, we predict that certain dipolar squares can transition between percolated and nematic states, depending on the system's density.

Our results highlight how colloidal particles with several degrees of anisotropy exhibit a rich phase behavior that sensitively depends on particle geometry and directional interactions. Our theoretical predictions are useful to colloidal scientists that are attempting to rationally synthesize particles with controllable properties. Finally, our studies of dipolar squares in the absence of a magnetic field provide a useful background as we extend our simulations to dipolar squares in the presence of an external magnetic field.

133a Nature-Derived Nanostructures for Sustainable Si Anode in High-Energy Li-Ion Batteries

Zheng Chen

Silicon (Si) has attracted considerable interests as a high-capacity anode material for next-generation lithium-ion batteries. However, Si-based anodes suffer extreme volume change (~380%) upon lithiation and delithiation, which results in rapid capacity fading due to mechanical and electrochemical failure during cycling. In this talk, we report a sustainable and scalable method to synthesize hierarchically porous micron-sized Si particles from the low-cost diatomite precursor, which serves as both the precursor and the template. Through a one-step magnesiothermic reduction, the SiO_2 constituent in diatomite is reduced to form a Si/SiO_2 composite network with 10–30 nm crystalline Si domains embedded within an amorphous SiO_2 matrix. Controlling the reduction time leads to an optimal ratio between the crystalline Si and the amorphous SiO_2 constituent, which endows the composite structure with high capacity and excellent cycling stability. For example, 90% capacity can be retained after 500 cycles at 0.2C for sample reduced by 6 h without any coating or prelithiation. The full cell with such Si/SiO_2 as the anode and $LiNi_{0.8}Co_{0.1}Mn_{0.1}O_2$ as the cathode showed ~80% capacity retention after 200 cycles. This work creates a unique path towards sustainable and scalable production of high-performance micron-sized Si anodes, offering new opportunities for potential industrial applications.

133b Rational Design of Interphases to Enable High Energy Metal-Based Batteries

Christopher Fetrow, Matthew Powell, Cameron Carugati, Shuya Wei

Advances in the basic science and engineering principles of electrochemical energy storage is imperative for significant progress in electronic devices. Metal-based batteries comprising of a metal (like Li, Na, Al, Zn) as the anode have attracted remarkable attention due to their promise of improving the anode-specific capacity by as much as 10-fold, compared to the current state-of-art Li-ion battery that uses a graphitic anode. The metal anodes also enable the utilization of energetic simple molecules (like sulfur, oxygen, carbon dioxide) as the cathode, which could further increase the cell level energy density. However, a persistent challenge with almost all metal batteries concerns their propensity to fail due to short-circuits produced by dendrite growth during battery recharge, as well as by runaway of the cell resistance due to internal side reactions with the liquid electrolyte. In this talk, I will discuss our research that utilizes ion transport modeling and contemporary experimental efforts to fundamentally understand and to thereby develop rational designs for electrode-electrolyte interphases that overcome these challenges for metal-based batteries. Particularly, on the anode side, we demonstrated that porous zinc electrodes reduce dendrite formation by lowering the diffusion-limited current density. To pair a metal anode with a small molecule (CO_2) gas cathode, we also demonstrated the design of the cathode and the electrolyte that enable a rechargeable Al-CO_2 battery.

133d Sequestration of Sulfur in Facilely Manufactured Carbon Nanospheres through Sulfur Recrystallization As an Effective PATH for LONG CYCLE Life of Lithium Sulfur Batteries

Wissam Fawaz, Zhao Wang, K.Y. Simon Ng

Lithium sulfur batteries are plagued by dissolution of sulfur into the electrolyte which gravely reduces their cycle life. Current strategies for retarding the negative fallout of the shuttle effect behavior experienced by lithium sulfur batteries, include attempts at trapping sulfur in various conductive carbonaceous matrices. In our work, we trap sulfur crystals in carbon nano-spheres through a facile synthesis technique to produces stable lithium sulfur cells with long lasting cycle life at C/10 rate. The trapping of the sulfur inside the sulfur nanospheres helps to delay the shuttle effect and protect the lithium anode from the negative effects of side reactions with sulfur polysulfides. Carbon nanospheres are constructed through a traditional carbonization of dextrose in the presence of dissolve sulfur. When the nanospheres are formed, sulfur is allowed to re-crystallize, and a coating of the finished material is utilized as the active cathode in the lithium sulfur cell. The formed material demonstrated its electrochemical suitability through studies of the electrochemical impedance of the cell as well as its cycle life. Formed cells show no demise of the high capacity of 850 mAh/g of sulfur over 150 cycles at a C/10 rate. We believe that confining the sulfur to nanospheres created in a sulfuric bath then allowing sulfur to recrystallize inside the sphere will slow down its dissolution into the polysulfide and extend the cycle life of the battery. The system will be optimized through varying the sulfur loading, crystal size, and carbon porosity as each variable contributes in our understanding of the stability displayed by the formed electrode

133e Investigating the Diffusivity of Ionic Liquids in Solvent-in-Salt System Using Molecular Dynamics Simulation

Ray Matsumoto, Wei Zhao, Xiaobo Lin, Ivan Popov, Alexei Sokolov, Peter Cummings

Highly concentrated salt electrolytes, known as solvent-in-salt (SIS) systems, have attracted significant attention recently as electrolytes for energy storage devices.[1–3] Due to the high cation transference number (the ratio of the diffusion coefficient of cation to the total diffusion coefficient of ions), wide range of electrochemical stability, suppression of lithium dendrite formation, low flammability, low volatility, and high charge density, SIS systems are now considered as promising electrolyte systems for the next generation batteries and supercapacitors. Compared to diluted electrolytes and pristine ionic liquids, it has been shown that the small amount of solvent makes dynamics and structure of SIS systems significantly different.[4,5]. However, the microscopic transport properties of ions in SIS systems remain poorly studied. Thus, understanding the factors that influence the charge transport in SIS systems is crucial for their design and applications.

In this presentation, we discuss our recent work[6] on SIS systems consisting of highly concentrated LiTFSI-acetonitrile, and LiTFSI-water, in which the ratio of ion to solvent is 1:3. Using a combination of broadband dielectric spectroscopy (BDS) and molecular simulation, we analyzed the role of ion-ion correlation, ion-solvent correlation, and solvent-solvent correlation on the ion transport properties. By implementing the autocorrelation function of electric current through the Green-Kubo relation, the conductivity of the solvent-in-salt system can be directly generated, and the different contributions from the correlations between different ions and solvent can be analyzed explicitly. Moreover, the effect of concentration and temperature on the conductivity were also examined. The results show that a decrease in anion-cation correlations leads to a higher transference number of lightest ion, and that the cation-anion correlations can be tuned by changing solvent concentration.

References

1. Li, M., Wang, C., Chen, Z., Xu, K. andamp; Lu, J. New Concepts in Electrolytes. *Chem. Rev.* **120**, 6783–6819 (2020).
2. Suo, L., Hu, Y.-S., Li, H., Armand, M. andamp; Chen, L. A new class of Solvent-in-Salt electrolyte for high-energy rechargeable metallic lithium batteries. *Nat. Commun.* **4**, 1481 (2013).
3. Azov, V. A., Egorova, K. S., Seitkalieva, M. M., Kashin, A. S. andamp; Ananikov, V. P. "Solvent-in-salt" systems for design of new materials in chemistry, biology and energy research. *Chem. Soc. Rev.* **47**, 1250–1284 (2018).
4. Seo, D. M., Borodin, O., Balogh, D., O'Connell, M., Ly, Q., Han, S.-D., Passerini, S. andamp; Henderson, W. A. Electrolyte Solvation and Ionic Association III. Acetonitrile-Lithium Salt Mixtures–Transport Properties. *J. Electrochem. Soc.* **160**, A1061–A1070 (2013).
5. Popov, I., Sacci, R. L., Sanders, N. C., Matsumoto, R. A., Thompson, M. W., Osti, N. C., Kobayashi, T., Tyagi, M., Mamontov, E., Pruski, M., Cummings, P. T. andamp; Sokolov, A. P. Critical Role of Anion–Solvent Interactions for Dynamics of Solvent-in-Salt Solutions. *J. Phys. Chem. C* **124**, 8457–8466 (2020).
6. Popov, I., Khamzin, A., Matsumoto, R. A., Zhao, W., Lin X., Cummings, P. T., Sokolov, A. P. Controlling ion transport number in Solvent-in-Salt Solutions. (Submitted)

133f Enhancing the Oxidative Stability of Mxenes

Xiaofei Zhao, Jodie Lutkenhaus, Miladin Radovic, Micah Green

The oxidation of 2D MXenes jeopardizes their shelf life, both in colloidal dispersions and in functional energy-storage devices. Our group has developed several approaches to prevent MXene oxidation, including low-temperature storage, antioxidants, and thermal annealing. Antioxidants in particular have been shown to effectively mitigate oxidation of MXenes (such as sodium L-ascorbate, ascorbic acid, and polyanions), but the nature of interaction between these antioxidants and MXene remains unknown, which impedes the future selection and design of improved protection. This work systematically examines the interactions between three classes of antioxidant candidates, α-hydroxy acids, polycarboxylic acids, and phenols with $Ti_{n+1}C_nT_x$ MXenes, specifically $Ti_3C_2T_x$ and Ti_2CT_x. Interestingly, while some antioxidants provided no protection for the MXenes, and some antioxidants even accelerated their degradation, three antioxidants (*e.g.,* citric acid, tartaric acid, and oxalic acid) protected the MXene nanosheets exceptionally well, showing minimum MXene degradation storage. Analysis of the antioxidants' molecular structure and efficacy suggests that chelation interactions with the transition metal atoms of the nanosheets play a key role in effective protection of MXenes from oxidation.

133g PEGDA-ZIF-8 Composite Electrolyte for All-Solid-State Lithium Metal Batteries

Zizhou He, Joshua Goulas, Rayden Farmer, Belle Racca, Ling Fei

Li metal is considered to be the most promising anode for high-energy batteries as it has a high specific capacity and low electrochemical potential. However, traditional liquid electrolytes can't meet the practical application requirement due to safety issues. Solid-state polymer electrolytes (SPE) with low flammability, wide electrochemical stability window, and sufficient mechanical stability have been developed to address the battery safety issues. Still, polymers tend to recrystallize at room temperature, leading to insufficient ionic conductivity. Metal-organic frameworks have been found to effectively increase the ionic conductivity of SPEs by decreasing the crystallinity polymer, immobilizing anions, and improving Li^+ transfer ability. Porous zeolitic imidazolate framework-8 (ZIF-8) has been used as fillers in this study. The ZIF-8 particles are intergraded with the Polyethylene Glycol Diacrylate (PEGDA) matrix with a simple UV-curing process in this study. The resulted composite electrolyte delivers a room Li^+ ion conductivity of 3.6×10^{-4} S cm^{-1} and outstanding electrochemical stability with a wide electrochemical window of 0–4.8 V (versus Li/Li^+). The good performance is due to the composite solid-state electrolyte inherited the advantages of both polymer and fillers. On one hand, the polymeric PEGDA enhanced the contact interphase between the electrolyte and electrodes. On the other hand, fillers can inhibit the recrystallization of the polymer, improve Li^+ transfer ability, as well as prevent dendrite growth. The simple fabrication process and good electrochemical performance of PEGDA-ZIF-8 composites are of great promise for all-solid-state lithium metal batteries.

202a Machine Learning Approach to First-Principles Database for Designing Active Nanomaterials for Electrochemical Energy Convergence

Byungchan Han, Hoje Chun, Minjoon Hong

Machine learning approach has been intensively applied to nanomaterials of high activity for energy convergence, due to its evident power to identify a key correlation in database. The success of the method, therefore, is critically dependent on the reliability and accuracy of accumulated data. Here, the acquisition of crystal-clear structure-performance relation in heterogeneous catalysis of sluggish electrochemical reactions is demonstrated through three-dimensional tracking of nanoparticles in liquid medium and over thermal treatment. For typical catalyst design process using colloidal synthesis we acquire atomic positions of ligand-protected platinum nanoparticles using first-principles density functional theory calculations and in-situ TEM observation. The structural transformation in time domain is shown by decoupling and tracking of each atomic coordinate using molecular dynamics simulations standing on machine-learning potentials. Fast and accurate acquisition of structural data is integrated with a computational catalysis framework to identify a clear structure-activity correlation in the CO oxidation reaction as an example.

202b Modulating the Active Sites of Metal–Nitrogen–Doped Carbon Catalysts By Orbital Coupling for Highly Active and Selective CO_2 Electrochemical Reduction

Jeong Woo Han

The electrochemical CO_2 reduction reaction (CO_2RR) is a potential strategy for converting waste CO_2 into a valuable chemical and closing the CO_2 cycle. We propose a computational screening route for highly active and selective metal-nitrogen-doped carbon (M-N-C) catalysts, followed by their experimental validations. This screening approach leads to a more reasonable electrocatalyst design as well as insights into the relationship between experimental and theoretical investigations. In addition, we extended our study of catalysts with dual atom centers beyond a single M-N-C. The neighboring iron dual-sites anchored on nitrogen-doped carbon matrix (Fe_2-N-C) outperform their single atomic counterparts in terms of CO Faradaic efficiency over wider potential ranges along with turnover frequency and durability. The orbital coupling between the iron dual-sites modifies the energy gap between antibonding and bonding states in CO binding, hence decreasing CO binding energy and promoting activity. This research presents new insights into the structure–performance relationship on CO_2RR electrocatalysts at the atomic scale and extends the application of dual-atom electrocatalysts for heterogeneous electrocatalysis and beyond.

202c *Sp-D* Orbital Hybridization Driven Metal-Graphene and Metal-Graphene-Metal Catalyst for Direct and Electrochemical Synthesis of Hcooh from CO_2: First-Principles Approach

Jinwon Cho, Ji Il Choi, Matthew Drexler, Faisal M. Alamgir, Seung Soon Jang

The efficient transformation of easily available CO_2 into renewable energy sources is a crucial for the decarbonization. Formic acid (FA), the simplest combination of hydrogen and CO_2, is one of the most promising hydrogen storage compounds for its high hydrogen density, non-toxicity, and great stability under ambient temperature and pressure. However, direct synthesis of CO_2 to FA on the noble metal and metal alloy catalysts has been a challenge for a decade since the reaction is thermodynamically unfavorable and kinetically difficult due to the nature of relatively weak adsorption of CO_2 on the metal surfaces in comparison to HCOO, and thus requires very high pressure/temperature. Electrochemical conversion of CO_2 on the conventional carbon black supported noble metal and metal alloy catalysts, in contrast, operates at ambient conditions, but suffer low stability through dissolution and agglomeration, and requires high overpotentials to overcome a competitive hydrogen evolution reaction (HER). In this context, we investigate a highly active and stable catalysts obtained by depositing a transition metal monolayer on graphene (M/Gr) using DFT method. It is found that the hybridization of *sp* and *d* orbitals between graphene and metal significantly reduces the local density of states of *d*-band of metals and increases *s*-and *p*-orbital of graphene near the fermi level at the same time, which leads to a strong covalent bonding between metal monolayer and graphene. Especially, the calculation of the orbital-resolved density of states for *d* band of M unravels that sharp d_{z^2}, d_{yz}, and d_{xz} orbital peaks below the fermi level play a decisive role in the formation of covalent bonding between M and Gr. In addition, it is found that the charge polarization on graphene in Pt/Gr and Pd/Gr not only enables a deposition of another metallic monolayer on the graphene, thus forming Pt/Gr/M or Pd/Gr/M sandwich structures, but also induces a transition from physisorption to chemisorption for M = Ag, Au, and Cu cases. Finally, it is discovered that the direct synthesis of FA from CO_2 is the most energetically favorable on Pt/Gr catalyst, and by evaluating the overpotentials required for the electrochemical CO_2 reduction to FA, among M/Gr, Pt/Gr/M, and Pd/Gr/M electrocatalysts Ni/Gr, Pd/Gr, Pt/Gr/Ag, and Pt/Gr have excellent activity and selectivity toward electrosynthesis of FA.

202d High-Temperature Flow Synthesis of Lead Halide Perovskite Nanocrystals

Kameel Abdel-Latif, Fazel Bateni, Mahdi Ramezani, Milad Abolhasani

Since their discovery in 2015, all-inorganic lead halide perovskite (LHP) nanocrystals (NCs) have proven an excellent candidate to outperform conventional II-VI semiconductor NCs. [1] The facile and broad emission tunability of LHP NCs coupled with their narrow emission linewidth and high quantum yield make them an ideal candidate for next-generation printed energy technologies.

The most commonly used approach for discovery, fundamental studies, and development of LHP NCs is through conventional batch methods (*e.g.*, round-bottom flasks). [2] Despite the ease of setup assembly, the batch reactors are suffering from limited heat and mass transport rates, absence of appropriate *in situ* materials characterization probes, batch-to-batch variations, and high reagents consumption. Recently, microfluidic platforms have gained significant interest for continuous manufacturing and precision synthesis of complex nanomaterials.[2] The reduced active reactor volume of microreactors compared to batch reactors allows for reliable nanomaterial synthesis and characterization with intensified transport phenomena, integration of *in situ* spectral monitoring probes, facile process automation, and reduced reagent consumption and waster generation. [2] These unique attributes make microfluidic platforms a highly effective tool for accelerated synthesis, optimization, and fundamental formation mechanism studies of high-quality LHP NCs. Despite the pioneering work in synthesis of LHP NCs, their formation mechanism using the widely adopted high-temperature synthetic route is not well-understood, thereby limiting the large scale production of LHP NCs for adoption by emerging printed energy technologies. [3]

In this work, we present, a modular and reconfigurable flow synthesis platform to study and unveil the formation mechanism of LHP NCs *via* one-pot high-temperature synthetic route. The unique modular design of the developed microfluidic platform allows access to three high-temperature synthetic routes of LHP NCs, including heat-up, hot-injection. Next, we utilized the developed flow synthesis platform to deconvolute the effects of interdependent reaction parameters controlling the NC morphology (nanocube *vs.* nanoplatelet) and optical properties of a model LHP NCs, cesium lead iodide ($CsPbI_3$). The results of the in-flow high-temperature synthesis of $CsPbI_3$ NCs, unveiled a change in the NC formation mechanism at temperatures higher than 150°C, manifested by the observed change in the NC morphology. Additionally, capitalizing on the excellent heat and mass transfer rates offered by microreactors, we demonstrated the unique capability of flow reactors for scalable production of high-quality $CsPbI_3$ NCs, independent of the target product volume.

The developed modular flow chemistry technology unveiled in this work provides a new frontier for high-temperature studies of solution-processed LHP NCs and enables their consistent and reliable continuous nanomanufacturing for next-generation energy technologies.

References:

[1] Abdel-Latif, K.; Bateni, F.; Crouse, S.; Abolhasani, M. Flow Synthesis of Metal Halide Perovskite Quantum Dots: From Rapid Parameter Space Mapping to AI-Guided Modular Manufacturing. *Matter* **2020**, *3* (4), 1053–1086. https://doi.org/10.1016/j.matt.2020.07.024.

[2] Abdel-Latif, K.; Epps, R. W.; Kerr, C. B.; Papa, C. M.; Castellano, F. N.; Abolhasani, M. Facile Room-Temperature Anion Exchange Reactions of Inorganic Perovskite Quantum Dots Enabled by a Modular Microfluidic Platform. *Adv. Funct. Mater.* **2019**, *29* (23), 1900712.

https://doi.org/10.1002/adfm.201900712.

[3] Antami, K.; Bateni, F.; Ramezani, M.; Hauke, C. E.; Castellano, F. N.; Abolhasani, M. $CsPbI_3$ Nanocrystals Go with the Flow: From Formation Mechanism to Continuous Nanomanufacturing. *Adv Funct Materials* **2021**, 2108687. https://doi.org/10.1002/adfm.202108687.

202e Photo-Catalyzed Polymerization of Substituted Anilines Via Psi Active Sites

Marc Nabhan, G. Kane Jennings, David Cliffel, Tyler Oddo

Photosystem I, an intrinsic protein complex in the light-depending reactions of oxygenic photosynthesis, has witnessed increasing applications in vitro for its high quantum yield and photoelectrochemical properties. The potential resulting from electron-hole splitting at the PSI active sites can induce a polymerization of substituted aniline monomers at the P_{700} active site. Polyaniline (PAni) is a conducting polymer that can be synthesized through oxidative polymerization of aniline. The P_{700} active site, with its 490 mV vs SHE reduction potential, should allow PSI to initiate and propagate oxidative polymerizations for a specific set of energetically aligned substituted aniline monomers. In this work, we report the photocatalyzed oxidation of para and ortho anisidine (methoxyaniline) via extracted PSI molecules in the presence of simulated sunlight. The emergence of a protein-polymer complex through the photo-induced polymerization is confirmed through infrared spectroscopy, UV-Vis spectroscopy, thermogravimetric analysis, and contact angle measurements. The level of conjugation and intermolecular attraction between the protein and the resulting polymer is investigated using analytical ultracentrifugation, SDS-PAGE, and photoelectrochemical measurements of protein-polymer films.

202f Kinetic Control of Intrinsic Pores in Monolayer Graphene for Large-Area Proton Selective Membranes

Piran Kidambi

Angstrom scale proton selective defects introduced into the lattice of atomically thin 2D materials such as graphene viascalable processes present potential for transformative advances as membranes for energy conversion/storageapplications. Here, we show for the first time that kinetic control of graphene synthesis during chemical vapordeposition (CVD) can allow for tunable Angstrom scale proton selective pores for facile fabrication of large-areaatomically thin proton exchange membranes. Using a resistance-based transport model in conjunction with systematicliquid and gas phase transport measurements on the same membrane, we study the influence of such Angstrom scaleintrinsic defects on selective proton transport through centimeter-scale Nafion|Graphene|Nafion sandwich membranes.Additionally, we report on a novel graphene stacking approach that effectively mitigates transport of un-desiredspecies(small ions and gas crossover)without adversely affecting proton transport. Our insights on kinetic control ofgraphene synthesis for Angstrom scale pore formation over large areas as well as facile stacking approach formembrane fabrication offer new avenues to enable functional large-area atomically thin proton exchange membranes.

330a Synthesis and Application of the Spherical Nanoporous Carbons for Energy Conversion and Energy Harvesting

Chanho Pak, Jong Gyeong Kim, Sunghoon Han, Jisue Kang, Seokjun Cha, Jong-Jin Park

Nanoporous carbons that have nanopores as pore size of nano dimensions have been widely adopted in energy-related applications such as supercapacitors, supports and catalysts for fuel cells due to the high surface areas, tunable average pores size, and particle shapes [1-4]. Among the many shapes of the nanoporous carbon, spherical nanoporous carbons (SNCs) prepared via hard-templated methods have several unique features such as tunable particle size distribution, regular morphology, and controllable surface areas [1,4]. In this presentation, the SNCs with Fe and N doping for oxygen reduction reaction (ORR) as a usual application is discussed at first. And the SNCs with different average sizes for the triboelectric nanogenerator (TENG) for energy harvesting of wasted mechanical energy as a new application area are presented [5,6].

As hard-template materials, mesoporous spherical silicas (MSSs) with different particle sizes were prepared by the Stöber process using high concentrations of tetraethyl orthosilicate (TEOS) and n - hexadecylamine as a nonionic surfactant. To control the particle size of MSS, the concentration of the ammonia is adjusted [5,6]. The Fe-SNC catalysts were prepared using the incipient wetness impregnation method with Iron(III) chloride hexahydrate and 1,10-phenanthroline in a mixture of ethanol and sulfuric acid. The four Fe-SNC catalysts are prepared using four different MSS with controlled particles sizes, which are designated as Fe-SNC_151, _337,_723, and _1173. The number at the end of the sample name is the average size of the catalyst in nm, respectively [5]. For the energy harvesting application, three SNC samples with sizes of 1000, 700, and 300 nm were prepared with phenanthrene, p-toluenesulfonic acid, and MSS as a hard template. The electrode for the TENG with SNC samples was prepared with PVDF nanofiber and gradient distribution of the different sizes of SNC samples [6].

As a result, among the Fe-SNC catalysts, the Fe-SNC_337 showed the highest half-wave potential and the second highest limiting current values; 0.918 V and 5.881 mA/cm^2, respectively, which is a higher activity than that of Pt/C in the alkaline conditions. The ORR activity was dependent on the catalyst particle size. This size effect can be explained by the fact that the smaller catalyst particles allow for improved electrolyte accessibility; this provides access to a larger number of active sites within the same reaction time. One exception to this was Fe-SNC_151, which has agglomerated particles, unnecessary active sites and a different limiting current because of the small protrusions on the catalyst surface due to undesired iron species. This resulted in low ORR performance for Fe-SNC_151, despite it having the smallest catalyst particle size.

In addition, the SNC samples with different three sizes of 1000, 700, and 300 nm, respectively, was adopted in the positive and negative electrode with gradient structure and displayed the enhanced the charge confinement in the smaller SNC sample by fast transferring the surface charge from the larger SNC sample. Via this process, the output voltage is changed from 15.2 V to 600 V after high voltage charge injection, thus representing an increase of about 40 times, which leads to turning on the 300 LEDs. Further, to amplify the low output current, which is a disadvantage of triboelectric energy, two types of electrical energy (triboelectric and electromagnetic energy) are produced in a single mechanical device. The output current produced by the cylindrical TENG and electromagnetic generator is recorded as being 1300 times higher, increasing from 12.8 μA to 17.5 mA.

To sum up, the hard-templated SNC samples can enhance the electrochemical reduction of oxygen if

Fe and N are introduced into the carbon matrix. A new application of SNC samples with different particles sizes is proposed in the energy harvesting of mechanical energy.

[1] J. Liu, N. P. Wickramaratne, S. Z. Qiao, M. Jaroniec, *Nat. Mater.*, **14**, 763 (2015).

[2] H. Shao, Y.-C. Wu, Z. Lin, P.-L. Taberna, P. Simon, *Chem. Soc. Rev.* **49**, 3005 (2020).

[3] V. Yarlagadda, N. Ramaswamy, R. S. Kukreja, S. Kumaraguru, *J. Power Sources*, **532**, 231349 (2022).

[4] H. Chang, S. H. Joo, C. Pak, *J. Mater. Chem.* **17**, 3078 (2007).

[5] J. Lee, J. G. Kim, C. Pak, *J. Energy Chem.* **52**, 326 (2021).

[6] S. Cha, Y. Cho, J. G. Kim, H. Choi, D. Ahn, K. Sun, D.-S. Kang, C. Pak, J.-J. Park, *Small Methods*, 2101545 (2022).

330b Microscopic Modeling Ligand Crosslinking in Nanopatterning of Quantum Dots (QDs)

Niranjan Sitapure, Tae Hyun Kwon, Jeehye Yang, Moon Sung Kang, Joseph Kwon

The utilization of quantum dots (QDs) for the development of high-resolution displays (HDRs) has garnered significant attention due to their high photoluminescence quantum yield, wide color gamut, tunable optoelectronic properties [1]. The development of HDRs requires precise nanopatterning of QDs to generate tiny RGB pixels, which is a fundamental unit in HDRs. Unfortunately, during repeated nanopatterning of QDs, the deposited QD films are damaged when exposed to various solvents in the post-processing steps [2,3]. Thus, the literature is rife with different approaches for altering the nanopatterning process and developing robust QD films. Amongst various approaches, crosslinking of QD surface ligands to form crosslinked QD network has shown great promise [4,5]. Specifically, such a crosslinked network interlocks the neighboring QDs and increases the QD thin-film resistance during post-processing. For this method, crosslinker structure, and the number of crosslinkers have significant effects on the crosslinking performance. However, experiments could not fully ascertain the mechanism of the crosslinking process [4]. To this end, a kinetic Monte Carlo (kMC) model is developed to elucidate the mechanism and kinetics of the crosslinking process. Specifically, the ligand crosslinking reaction is broken down into two sub-steps (i.e., radical formation, and a C-H insertion reaction). Then, the surface reaction kinetics for these two sub-steps is modeled and integrated with a 2D kMC lattice. In this model, different spatial crosslinking configurations between the crosslinkers and ligands are reflected considering the geometry, dimensions, and structure of the crosslinkers. For example, the crosslinking efficiency of planar 4-armed crosslinkers is found to be higher with tetrahedral 2-armed crosslinkers because of their structural compatibility with the surface-ligand shell and the two additional crosslinking arms. Overall, a) the developed kMC model virtually probes into the crosslinking mechanism; b) helps quantify the crosslinking performance of different crosslinkers, and c) the simulation results are in excellent agreement with the experimental observations.

References:

1. Protesescu, Loredana, et al. 'Nanocrystals of cesium lead halide perovskites (CsPbX3, X= Cl, Br, and I): novel optoelectronic materials showing bright emission with wide color gamut.' Nano letters 15.6 (2015): 3692-3696.
2. Liu, Guiju, et al. 'Stable tandem luminescent solar concentrators based on CdSe/CdS quantum dots and carbon dots.' Journal of Materials Chemistry C 6.37 (2018): 10059-10066.
3. Wang, Li, et al. 'Giant stability enhancement of CsPbX3 nanocrystal films by plasma-induced ligand polymerization.' ACS applied materials andamp; interfaces 11.38 (2019): 35270-35276.
4. Yang, Jeehye, et al. 'High-resolution patterning of colloidal quantum dots via non-destructive, light-driven ligand crosslinking.' Nature communications 11.1 (2020): 1-9.
5. Ahn, Junhyuk, et al. 'Ink-Lithography for Property Engineering and Patterning of Nanocrystal Thin Films.' ACS nano 15.10 (2021): 15667-15675.

330c On the Importance of the Electric Double Layer Structure in Electrocatalysis

Hyungjun Kim

Electrochemistry, the fundamental basis of sustainable energy conversion technologies, studies the interconversion of electric-chemical energy at heterogeneous interfaces. The electrochemical reaction mainly occurs at the solid-liquid (or electrode-electrolyte) interface, where a characteristic liquid structure, i.e., an electric double layer (EDL) is formed. Since the early 1900s, when the concept of EDL was first proposed, unremitting efforts have been made to identify EDL structural changes depending on the applied potential, but only a few molecular details have been disclosed to date. One famous example is EDL capacitance, an indicative quantity of the EDL structural change, which shows two characteristic peaks in a dilute electrolyte, but no molecular theory of liquid structure has fully explained them. To address this century-long debate, we develop a molecular simulation method to model the electrified interfaces based on quantum mechanical electronic structure theory and energetics – a density functional theory in classical explicit solvents (DFT-CES). Using DFT-CES, we find unprecedented liquid structural changes and phase transitions of the EDL, which originate two capacitance peaks at the same potentials to the experiment. Furthermore, we envisage a new perspective for developing better electrochemical systems by tailoring the interface; there's plenty of room in between.

330e Tailored Mesoporous Structures of Lignin-Derived Nano-Carbons for High Performance Supercapacitors

Lu Yu, David Keffer, David Harper

As a natural polymer with a high carbon to oxygen ratio, lignin is often considered a waste byproduct of pulp and biofuel production. Lignin's random and complex heterogeneous structure poses a challenge to developing high-performance carbon materials. Nevertheless, previous work demonstrated promise using kraft lignin as a feedstock for carbon composites applications. This work presents an efficient one-step method to create highly mesoporous activated carbons from Kraft lignin. We achieved surface areas greater than 3,000 m^2 g^{-1} and a mesopore ratio of 80% to develop electrodes in high-performance supercapacitors. For comparison, we also used a traditional two-step chemical activation method to produce activated carbons to elucidate the precursor-process-structure-performance relationship of lignin feedstocks resulting in enhanced electrochemical performance. One-step activation produced ACs with ultrahigh surface area and high mesopore ratio, leading to ultrahigh capacitance and energy density of the fabricated SCs. In comparison ACs, produced through two-step activation, have limited surface area but high oxygen content. The hydrophilicity of ACs greatly decreases the internal resistance and benefits the capacitance even at high current density, thus achieving high power density of the fabricated SCs. Fourier-transform ion cyclotron resonance mass spectrometry (FT-ICR MS) combined with nuclear magnetic resonance (NMR) spectroscopy elucidated a clear understanding of lignin chemical structure. X-ray diffraction, scanning electron microscopy, X-ray photon spectroscopy and Raman spectroscopy were applied to characterize the disordered graphitic structures in the carbon composites. A techno-economic analysis confirms the feasibility of commercializing these inexpensive lignin-based porous carbons described in this work and demonstrates competitive renewable alternatives to commercially available activated carbons.

330f Heteroatom Doping in N-Coordinated Metal Site Embedded Graphene for Highly Active Oxygen Reduction Reaction

Ara Cho, Jeong Woo Han

Metal-N-doped carbon (M-N-C) has been in the spotlight as a potential cost-effective catalyst for oxygen reduction reaction (ORR). To go beyond the state-of-the-art Pt/C, a new design strategy to modulating the electronic structure of the material is necessary. Introducing a heteroatom around the metal center could influence the binding strength of the ORR intermediates by modifying the electronic structure of the metal atom, which could improve the ORR activity. Although several studies on the metal site coordinated by dual heteroatom have been conducted for ORR, systematic investigations depending on the type of heteroatom are still poorly understood. In this study, we investigated the ORR activities and electronic structural changes on metal active site for a total of 24 structures of $M-N_3Y$ (M=Fe, Ni, Co, Mn; Y=B, C, N, O, P, S) based on density functional theory (DFT) calculations. In addition, we investigated the effect of heteroatom doping on the ORR mechanism through microkinetic analysis. As the potential increased, the ORR intermediate poised the active point and affected the adsorption strength. Our findings will provide a useful guidance for designing novel ORR catalysts.

330g Polymer Threaded Metal Organic Frameworks with Ionic and Electrical Conductivity for Electrochemical Applications

Chi-Ying Vanessa Li, Ching-Kit Ho, Liang Gao

We discuss the recent progress made in modifying metal organic frameworks (MOF) to render ionic and electrical conductivity and enable their use in electrodes and membranes. When threaded with polyelectrolytes, MOF can be ion-exchangeable with high capacity [1, 2] as well as ionic conducting with high selectivity [3].

Poly(sodium vinyl sulfonated-co-acrylic acid)~MIL-53(Al) is prepared on anodic aluminum oxide substrate via steps of MOF MIL-53(Al) growth followed by in-situ polymerization [3]. The poly(VS-co-AA)~MIL-53(Al) membrane demonstrates highly specific selectivity in transport of alkali metal cations. Rate of ion transport correlates inversely with the hydrated diameter of the ion reaching a low limiting rate near 0.7 nm hydrated diameter. Charge exclusion is demonstrated with blockage of anion transport under a concentration gradient. The highly uniform porous nanostructure of MOF and ionic function of polyelectrolyte offers the MOF membrane with synergistic selectivity based on exclusion forces of the framework and Coulomb forces from fixed charges of polyelectrolytes in nanochannels.

The latest example is the electrical conducting polyaniline threaded MIL-101 and its application as a electrode material[4]. The combination of a 0.01 S cm^{-1} conductivity with ultrahigh surface area of 2065 m^2 g^{-1} offers unique opportunity for application in sensing and energy conversion. The Nernstian-like current response of the Fe^{2+}/Fe^{3+} redox couple indicates very rapid transport in the polymer thread MOF electrode, while the high surface area provides orders of magnitude higher current compared to ultramicroelectrode. The potential use of these electrodes in high current applications such as redox flow batteries is discussed.

Acknowledgement

The work described was fully supported by a grant from the Research Grants Council of the Hong Kong Special Administrative Region, China (Project No. T23-601/17-R).

[1] L. Gao, C.Y. V. Li, K.Y. Chan, and Z.N. Chen, *J. Am. Chem. Soc.* 136 (2014) 7209-7212.

[2] L. Gao, C.Y. V. Li, and K.Y. Chan, *Chem. Mater.* 27(10) (2015) 3601-3608.

[3] L. Gao, K.-Y. Chan, C.-Y. V. Li, L. Xie, and J. F. Olorunyomi, *Nano Lett.* 2019, doi.org/10.1021/acs.nanolett.9b01211.

[4] C.K. Ho, C.V.Y. Li, L. Gao, K.Y. Chan, J. Chen, J. Tang, J.F. Oloruyomi, C. Liao, and T.S. Zhao, *ACS Energy Lett.* 2021, 6, 11, 3769–3779.

Figure 1 Cyclic voltammetry curves of PANI~MIL-101 showing stable Fe^{2+}/Fe^{3+} ion selective activity for the first 50 cycles using 0.01 M aqueous $FeCl_2/FeCl_3$ in 1 M HCl, at a scan rate of 50 mV/s.

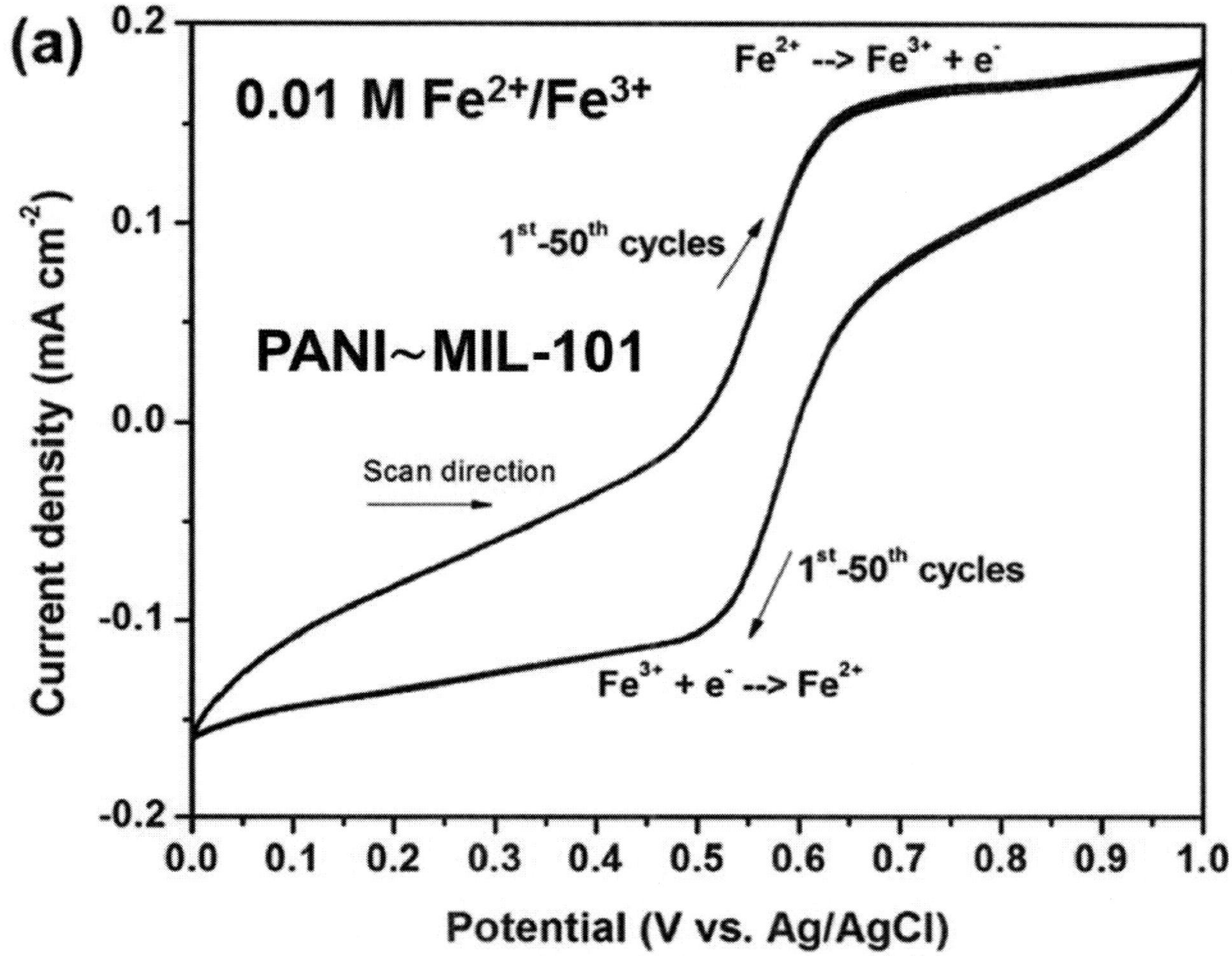
(a)
0.01 M Fe2+/Fe3+
PANI~MIL-101
Fe2+ --> Fe3+ + e-
1st-50th cycles
Scan direction
1st-50th cycles
Fe3+ + e- --> Fe2+
Current density (mA cm-2)
Potential (V vs. Ag/AgCl)
0.2
0.1
0.0
-0.1
-0.2
0.0
0.1
0.2
0.3
0.4
0.5
0.6
0.7
0.8
0.9
1.0

132a Investigating Serrs of Electrogenic Bacteria Via Charge Transfer Enhancement By Graphene Nano-Dots (GNDs)

Sheldon Cotts, Bijentimala Keisham, Roshan Nemade, Angelo Giles, Vikas Berry

Enhancement of bacterial peaks via Raman Spectroscopy has provided an efficient method of bacterial detection without lengthy cell cultures. We report the use of graphene nano-dots (GNDs) to chemically detect and enhance select bacterial peaks by a factor of ten. This demonstrates the charge transfer ability of electroactive bacteria and provides a platform for discussing chemical enhancement phenomena for biological sensing at the cellular level. By integrating the molecule energy levels of an analyte with cytoelectrons (4.66-4.74 eV) and a GND (~4.9 eV) substrate, we can design new cytobotic constructs leveraging cellular respiration beyond standard electrochemical devices.

132b Adsorption of Methylene Blue from Aqueous Solution By Zirconia Nanoparticles on Multiwall Carbon Nanotubes

Julian Lopez, Javier Lara Romero, Alexis Pérez Gasquez y Marín, Andrés Alejandro Damian Reyna

Synthetic dyes are widely used as coloring agents in textile, paper, leather, gasoline, pharmaceutical, and food industries. These dyes interfere in the processes of aquatic life, preventing the free passage of light and affecting the operation of wastewater treatment plants. The dye degradation products are toxic, carcinogenic and cause an increase in the levels of chemical oxygen demand (COD) and biological oxygen demand (BOD).

In this work, we report the adsorption of methylene blue onto zirconia nanoparticles supported on multiwalled carbon nanotubes (MWCNTs) via a microwave-assisted method. For the synthesis of circonia nanoparticles, zirconium (IV) oxynitrate hydrate was used as a metal source. Dioctyl sulfosuccinate sodium salt and sodium borohydride were used as surfactant and reducing agent, respectively. (+)-α-Pinene and ferrocene were used as a carbon source and catalyst respectively for the synthesis of MWCNTs.

The resulting composites were characterized by HRTEM, XRD, TGA and Raman spectroscopy. This method enabled the growth of a homogeneous coating around the multiwall carbon nanotubes. The adsorption studies were carried out using UV-Vis spectroscopy. The produced nanocomposites showed higher dye adsorption capability and removal percentage (77.0 mg/g, 97.0 %) than the functionalized carbon nanotubes (37.1 mg/g, 56.7%). Increasing the temperature increased the adsorption capacity of the composite. The zirconia/multiwalled carbon nanotube composites show great potential to remove methylene blue from contaminated water.

132c Graphene and MOS2 GEL for Aligned 3D Printing for Electronic and Mechanical Study

Deisy Carvalho Fernandes, Vikas Berry, Philippe Poulin

Designing 3D printed micro-architectures using electronic materials with well-defined electronic structures will potentially lead to accessible device fabrication for 'on-demand' applications. Here we show gelation of 2D nanomaterial mixtures, their 3D gel-printing, and the resultant device's characterization for six different concentration of MoS_2/graphene (MoS_2:rGO) (100:0, 10:90,30:70, 50:50, 70:30, 90:10) and graphene/BN (BN:rGO) (50:50) alloys. Rheology studies show the viscoelastic properties of the gels, including viscosity and yield stress. The shear-thinning behavior was engineered to provide the ability of fabricate 3D printed device via an aligned controlled nozzle-extrusion based modified 3D printer, enabling the fabrication of composite electrodes. The 3D printed devices structure was arrested by lyophilization induced water removal. These electronic, structural and mechanical properties were investigated. The ordered graphitic region with sp^2 hybridized carbon atoms in the graphene sheets is in the order of 8.28 ±0.23 nm. The temperature-dependent electronic transport of 3D printed electrodes structures exhibits a transport-barrier of 16.24 meV and a tunneling width of 0.45 nm (Fowler Nordheim electron tunneling) with graphene centers having a carrier concentration of $1.7x10^{11}/cm^{-2}$. By modifying the graphene/MoS_2 concentration, the Young's Modulus of the structure can be controlled between 6 and 16 kPa. We envision that the proposed 3D-printing of gels of nanomaterial will lead to an evolution in the design of next-generation of 'on-demand' printed electronic and electromechanical devices.

201f Synthesis and Application of Glycopolymer-Wrapped Carbon Nanotubes for Detecting Carbohydrate-Protein Interactions

Ana DiLillo, Ka Keung Chan, Xue-Long Sun, Geyou Ao

The complexation of nanomaterials with biomolecules has been widely utilized for developing nano-enabled applications in biosensing, bioimaging, drug delivery, and therapeutics due to their exceptional optical, mechanical, and physiochemical properties. Synthetic glycopolymers – polymers with carbohydrate pendant groups – have been used extensively to mimic the functions of naturally occurring glycoconjugates. We created water soluble glycopolymer-wrapped single-wall carbon nanotubes (Glyco-SWCNTs) *via* noncovalent complexation utilizing lactose-containing polymers with various carbohydrate densities and chain lengths. The wrapping conformation of glycopolymers on the surface of SWCNTs enables a distinct nanostructure-induced interaction between the internal sugar group of glycopolymers and a specific carbohydrate-binding protein (CBP), as compared to currently known interaction behavior of CBPs with the terminal glycan of oligosaccharides. These Glyco-SWCNT complexes can potentially lead to the creation of versatile optical sensors for detecting carbohydrate-protein interactions with enhanced specificity and sensitivity.

510a Detection and Classification of Chiral Inorganic Particles in Electron Microscopy Images Using Generalizable Deep Learning Algorithms

Anastasiia Visheratina, Alexander Visheratin, Prashant Kumar, Michael Veksler, Nicholas Kotov

A chiral object has two mirror-image forms that are non-superimposable in three dimensions. Chirality plays a crucial role in chemistry, biology, and pharmacology, as most of the important biomolecules are chiral (amino acids, proteins, DNA). In 1998, it was discovered that chiral nanostructures could be chiral. Chiral inorganic nanostructures have distinct fundamental importance and are essential for further developing chemical, pharmaceutical, environmental, and biomedical technologies. To date, many researchers are focused on the establishment of the correlations between chiroptical and morphological properties of these materials by using circular dichroism and electron microscopies. A thorough investigation of electron microscopy images requires accumulating many images with their in-depth analysis, which is tedious and the 'manual' image processing is subject to experimentalist bias. Currently, there is a need for novel methods for the structural characterization of chiral nano- and micron-scale inorganic structures.

Here we developed an approach for synthesizing large sets of realistic scanning electron microscopy (SEM) images of chiral microscale helices of bowtie shape based on very small original SEM image sets (~200 SEM images). We used SEM images of a diverse pool of bowties, which allowed the training of state-of-the-art neural network models with the ability to detect morphological properties of chiral structures of various sizes. We tested the method by generating 10,000 images of bowtie-shaped particles with different chirality and training the YOLOv5 model to differentiate between right and left structures. As a result, this algorithm can reliably identify and classify chiral bowtie-shaped particles with accuracy as high as 94.4%. Furthermore, after training on bowtie particles, this model can successfully recognize *other* chiral shapes with different geometries without re-training. These findings indicate that deep learning techniques can potentially replicate the visual analysis of chiral objects, which opens up a path to other computational methods capable of accurate automated analysis of a wide range of chiral features at different scales and their implementation in materials discovery for photonics and medicine.

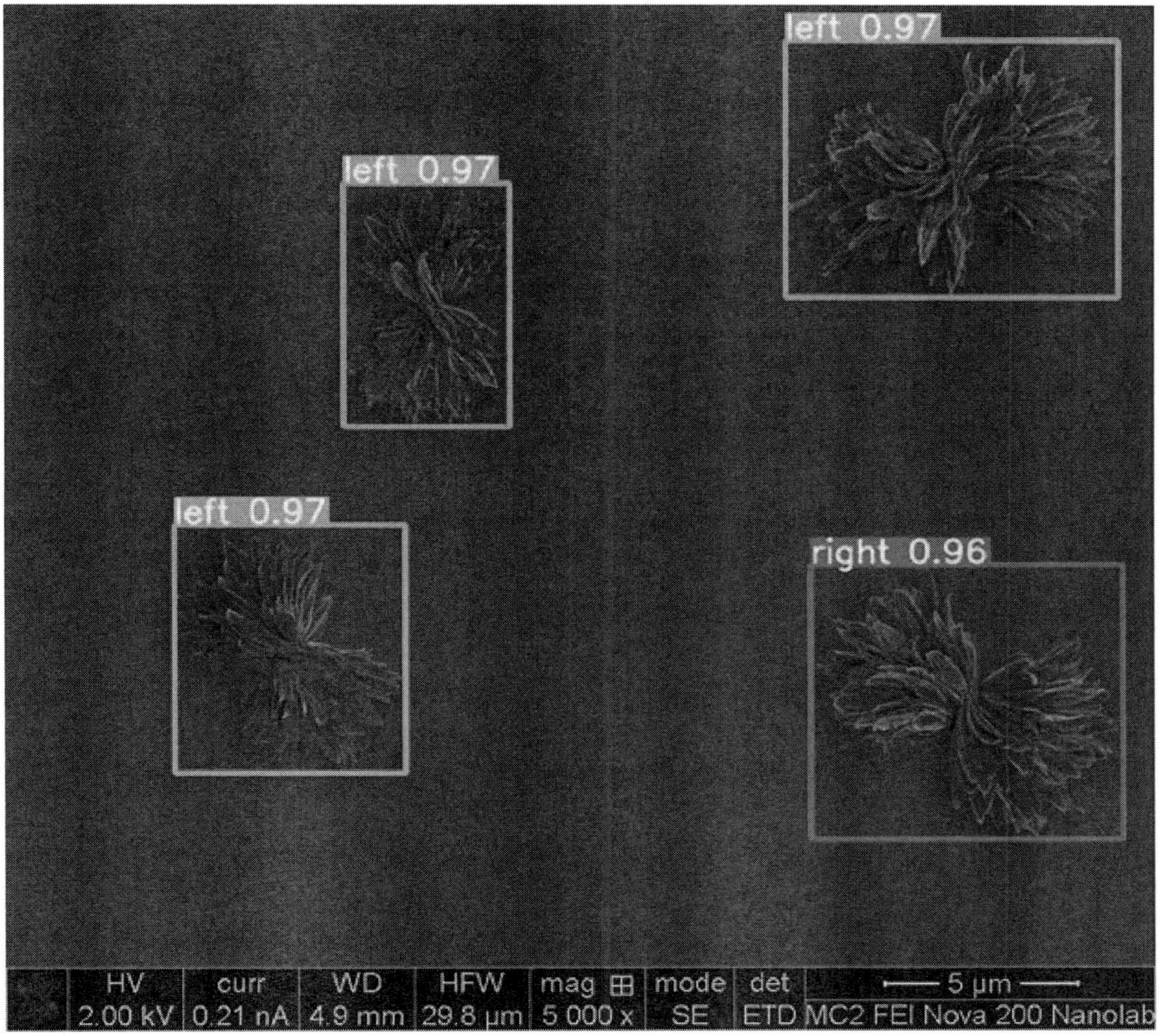

left 0.97
left 0.97
left 0.97
right 0.96
HV
2.00 kV
curr
0.21 nA
WD
4.9 mm
HFW
29.8 µm
mag
5 000 x
mode
SE
det
ETD
5 µm
MC2 FEI Nova 200 Nanolab

510b Association Rule Mining of the Relationships Among Biological Responses of Embryonic Zebrafish Exposed to Nanoparticles

Bilal Khan, Yoram Cohen

Relationships among the biological responses of embryonic zebrafish (including mortality) exposed to seven different types of nanoparticles at multiple concentrations were analyzed using association rule mining and clustering (hierarchical clustering and self-organizing maps (SOM)) analysis. Biological response data from a database on nanomaterial biological interactions (NBI) were extracted and used for clustering analysis and subsequently processed (categorized) for the application of association rule mining. The dataset consisted of ENMs characteristics, experimental conditions, ENMs concentrations, and effects of nanoparticles on embryonic zebrafish. Association rule mining was then applied to the biological responses dataset, followed by the identification of non-redundant rules. The non-redundant rules indicated that a significant impact of ENMs exposure to zebrafish on one or more biological responses implies significant impact of ENMs exposure to zebrafish on other (associated) responses. Pearson's pairwise correlation analysis, along with hierarchical clustering and SOM analysis indicated that the non-redundant association rules were consistent with confirmatory tests. The experimental results confirmed that zebrafish olfactory regions are inter-associated and heart is inter-associated with several responses especially touch response, circulation, delayed development, yolk sac edema, curved body axis, caudal fin, and pectoral fin deformations. Spontaneous zebrafish movements also correlated with yolk sac edema, axis and somites. The results of the current study suggest that there is merit in exploring the utility of the approach to streamline multiparametric biological responses for improved understanding of their relationships for the development of predictive toxicity nano-structure activity relationships.

510d Evaluation of Nucleation and Growth Kinetics of Ionic Liquid-Based Pt Nanoparticles Synthesis in a Millifluidic Reactor By in Situ Small-Angle X-Ray Scattering

Majed Madani, Noah Malmstadt

Ionic liquids have gained research interest as an alternative reaction medium for nanoparticle synthesis due to their ability to colloidally stabilize nanoparticles while inducing high nucleation rates. In contrast to organic solvents, which tend to be volatile and flammable, ionic liquids are environmentally friendly, and safe. Synthesis of Pt nanoparticles by polyol reduction in 1-butyl-3-methylimidazolium bis(trifluoromethylsulfonyl)imide (BMIM-Tf_2N) ionic liquid solvent has been reported as a viable approach to form colloidally stable Pt nanoparticles. However, the mechanism behind the nucleation and growth kinetics has not yet been explored. *In-Situ* Small-Angle X-ray Scattering (SAXS) provides the means to probe the formation mechanism in real-time. Herein, by utilizing SAXS, we present the real-time monitoring for the nucleation and growth kinetics of BMIM-Tf_2N-based Pt nanoparticles in a millifluidic reactor. We propose a robust and cost-effective reactor design for *in situ* SAXS measurements that allows for time resolution ranging from milliseconds to several minutes while operating at moderately high temperatures up to 350 °C. In this design, multiple measurement points are taken along the reactor channels, each corresponding to a specific time resolution along the reaction at steady state.

510e A First Step Towards the Development of a Nano-Production Line Using Multidisciplinary Quality By Design Approaches

Ramona Jeitler, Carolin Tetyczka, Christina Glader, Bianca Brandl, Daniela Fiedler, Eva Roblegg

The increasing molecular understanding of diseases, scientific advances in the development of large molecules, such as proteins, monoclonal antibodies and nucleic acids, and the global Corona pandemic have brought nanomedicine into the focus of research. Currently, fabrication of nano-delivery systems is a multi-step process using either non-continuous bulk processes or (parallelized) microfluidic technologies [1]. The latter include mixing, ultrafiltration to remove organic solvents, buffer exchange, and dilution to initial and final concentration, followed by bioburden filtration, sterile filtration and filling [2]. The smallest quality changes during such a multistage manufacturing process can lead to massive changes that impair efficacy and application. Despite enormous progress in research, industrial production processes are lagging behind, yet must not become a bottleneck in the future in order to keep pace with medical progress and ensure the best medical treatment for patients. Accordingly, there is a huge strategic need for the development of robust, scalable and versatile nano-manufacturing platforms that enable continuous, fast, high-yield, controllable and flexibly configurable production of nanomedicines at the large but also at the small scale.

This study presents a first step towards the development of a continuous and flexible nano-production line for reproducible production of nanoparticles with defined properties using multidisciplinary Quality by Design (QbD). The objective was to systematically identify and predict the ideal input parameters (i.e., process conditions and formulation composition) to optimize the size, size distribution, zeta potential, morphology, shape, internal structure, etc. of lipid nanoparticles according to regulatory guidelines (i.e., ICH Q8 (R2) and FDA-2017-D-0759). Design of Experiments (DoE) studies were conducted for this purpose.

To this end, drug-free solid lipid nanoparticles (SLN) consisting of Tween® 80 as stabilizer and one solid lipid respectively, i.e., either Palmitic acid, Compritol® 888 ATO or Geleol™ Mono- and Diglycerides NF were studied. In addition, more complex nanostructured lipid carriers (NLC) comprising one of the solid lipids and one of the liquid lipids, i.e., oleic acid, linoleic acid or medium-chain triglycerides stabilized with Tween® 80 were investigated. Both lipid formulations were dispersed by high-shear mixing (HSM) and further processed with high-pressure homogenization (HPH). Central Composite Face-centered (CCF) quadratic experimental design with multiple linear regression (MLR) were chosen to investigate the main, interaction, and quadratic effects of process conditions and formulation characteristics on SLN and NLC properties. After the initial risk assessment (RA), the total lipid concentration (i.e., 10 % w/w), batch size (i.e., 75 g) and process temperatures (i.e., 10°C above the melting point of the materials used) were set as fixed parameters. In contrast, the stabilizer concentration (i.e., 1-4 % w/w), the HSM speed (i.e., 8000-16000 rpm) and time (i.e., 15-60 s) as well as the HPH pressure (i.e., 250-750 bar) and the number of cycles (i.e., 1-10 (SLN) and 3-10 (NLC)) were considered as input in the DoE. In addition, for NLC preparation, different ratios of solid to liquid lipids (i.e., 9:1, 8:2, and 7:3) were tested. As suggested by MODDE®, a total of 29 experiments (26 runs with 3 center points to access the model reproducibility) were conducted for each formulation (i.e., 12). The particle size (i.e., d10, d50 and d90) and size distribution (i.e., span-value) was selected as response using Laser diffraction for characterization. For the statistical analysis the MODDE® software was used and coefficient plots, a summary of fit considering R2 (i.e., coefficient of determination), Q2 (i.e., goodness of prediction), model validity and reproducibility were created.

After the completion of the experimental data, the models were simplified by excluding small and insignificant interaction terms. Obtained coefficients plots for SLN formulations identified the applied number of cycles and the stabilizer concentration as well as their quadratic effects as the most significant terms of the model. In addition, interaction effects of pressure and number of cycles and number of cycles and stabilizer concentration show a minor effect on particle size and size distribution. The same was observed for the NLC formulations. In addition, the matrix composition (i.e., lipid ratio) and its quadratic effect also proved to have a significant influence. The summary of fit for SLN and NLC showed high R2-values (i.e., 0.803 and 0.913), which indicate high fit between data and model and high Q2-values (i.e., 0.595 and 0.850), which represent reliability of the prediction. Since R2 and Q2 were > 0.5 and differences between R2 and Q2 were < 0.3, the obtained models are considered highly valid. Moreover, via center point experiments in triplicate, reproducibility of the models was proven to be high (i.e., 1).

In summary, gaining proper process knowledge of the individual processes involved and a sound understanding of the individual formulation components in terms of processability and product quality can help bridge the gap between academic and batch-based research and industrially relevant large-scale production. Furthermore, this knowledge can be used to flexibly design continuous processes and monitor and control the critical process parameters and formulation properties during production using process analytical technologies (PAT) and control software.

References:

1. Sheperd, S.J. Microfluidic formulation of nanoparticles for biomedical applications, Biomaterials 274, 120826 (2021).
2. European medicines agency, Assesment report Onpattro, EMA/554262/2018 (2018).

510f Surface Mineralization of Barley Stripe Mosaic Virus Biotemplates

Che-Yu Chou, Yu-Hsuan Lee, Kok Zhi Lee, Chamath Vindula Basnayake Pussepitiyalage, Akash Vaidya, Shohreh Hemmati, Kevin Solomon, Sue Loesch-Fries, Michael T. Harris

Nanoscale materials are of a considerable interest in applications such as catalysis, sensing, and electronics. Biotemplates featuring tunable functional biomolecules provide hierarchical, controlled patterned nanostructure of high precision. The use of viral protein as biotemplate offers diversity of architecture with genetic/chemical tunability which creates morphology of precise dimension. Barley stripe mosaic virus virus-like particle (BSMV-VLP) has been designed and engineered to overcome the limitations related to viral genomic replication and plant-based production. The BSMV-VLPs produced via bacterial-based expression platform facilitate processing cycle for large-scale production. Uniform palladium coatings were synthesized on the BSMV-VLPs without an additional reducing agent. Our recent focus is to investigate the surface mineralization of BSMV virion produced from infected barley plants. The hydrothermal synthesis will be employed on BSMV and BSMV-VLP to create uniform and dense monometallic and bimetallic coatings. The synthesized metallic nanorods were imaged with transmission electron microscope (TEM). Scanning transmission electron microscopy (STEM), energy dispersive X-ray spectroscopy (EDS) and X-ray photoelectron spectroscopy (XPS) were used for elemental composition characterization. We have shown an effective and tunable approach for metallic nanorod deposition on biotemplates produced from different routes using hydrothermal solution synthesis. In summary, this synthetic approach has expanded the synthetic toolbox and broadened the possibilities in bottom-up processing of versatile and robust nanostructure and would expedite future industrial applications.

510g *In-Vivo* Fluorescent Nanosensor Implants Based on Hydrogel-Encapsulation: Minimization of Inflammation and the Foreign-Body Response

Michael A. Lee, Xiaojia Jin, Sureshkumar Muthupalani, Naveed Bakh, Michael Strano

Nanosensors are promising new tools for various *in vivo* applications such as sensing, delivery, and imaging. However, a wide range of tissue responses have been reported depending on cell type and various nanocomponent properties. The tissue response is critical in determining the health of the organism and the functional lifetime of the material. While nanomaterial properties can contribute largely to the tissue response, it may be possible to circumvent adverse reactions by formulation of the encapsulation vehicle. In this study, five formulations of poly(ethylene glycol) diacrylate (PEG-DA) hydrogel-encapsulated fluorescent nanosensors were implanted into mice, and inflammatory responses were tracked in order to determine the favorable design rules for hydrogel encapsulation. Five different immunocompromised mice lines were utilized for a comparison across different inflammatory cell populations and responses. Degradation products of the gels were also characterized. Finally, the importance of the tissue response in determining functional lifetime was demonstrated by measuring the time-dependent deactivation following implantation.

510h 3D-Printed Graphene/Polymer Structures for Electron-Tunneling Based Devices

Deisy Carvalho Fernandes, Vikas Berry, Philippe Poulin

Designing 3D printed micro-architectures using electronic materials with well-defined electronic structures will potentially lead to accessible device fabrication for 'on-demand' applications. Here we show gelation of 2D nanomaterial mixtures, their 3D gel-printing, and the resultant device's characterization for six different concentration of MoS_2/graphene (MoS_2:rGO) (100:0, 10:90,30:70, 50:50, 70:30, 90:10) and graphene/BN (BN:rGO) (50:50) alloys. Rheology studies show the viscoelastic properties of the gels, including viscosity and yield stress. The shear-thinning behavior was engineered to provide the ability of fabricate 3D printed device via an aligned controlled nozzle-extrusion based modified 3D printer, enabling the fabrication of composite electrodes. The 3D printed devices structure was arrested by lyophilization induced water removal. These electronic, structural and mechanical properties were investigated. The ordered graphitic region with sp^2 hybridized carbon atoms in the graphene sheets is in the order of 8.28 ±0.23 nm. The temperature-dependent electronic transport of 3D printed electrodes structures exhibits a transport-barrier of 16.24 meV and a tunneling width of 0.45 nm (Fowler Nordheim electron tunneling) with graphene centers having a carrier concentration of $1.7x10^{11}/cm^{-2}$. By modifying the graphene/MoS_2 concentration, the Young's Modulus of the structure can be controlled between 6 and 16 kPa. We envision that the proposed 3D-printing of gels of nanomaterial will lead to an evolution in the design of next-generation of 'on-demand' printed electronic and electromechanical devices.

331a Identifying Protein Corona Composition on Lipid Nanoparticle-Based mRNA Delivery Vehicles

Elizabeth Voke, Rebecca L. Pinals, Mariah Arral, Kathryn Whitehead, Markita Landry

Lipid nanoparticles are the most clinically advanced delivery vehicles for nucleic acid-based therapeutics, as demonstrated by the mRNA-based vaccines against SARS-CoV-2 from Moderna and Pfizer/BioNTECH. Applications of mRNA delivery expand beyond vaccines to include genome editing, protein replacement therapy, and immunotherapy. These lipid nanoparticles function to protect mRNA against degradation during transit and facilitate endosomal escape for the efficient delivery of their mRNA cargo upon cell internalization. However, nanoparticles encounter various biological tissues and compartments upon injection and biomolecules such as proteins spontaneously adsorb to the nanoparticle surface, coating the lipid nanoparticles in a protein "corona". Currently, we do not understand the connection between the protein corona formed and the delivery outcomes observed for lipid nanoparticles. Moreover, most studies of protein corona formation to date have employed dense nanoparticles with rigid structures (such as metal- or polystyrene-based nanoparticles), yet we expect soft nanoparticles will be the standard for mRNA-based delivery systems. As such, a better understanding of these biomolecular interactions on the surface of lipid nanoparticles may be critical to move beyond conventional outcomes of intravenous injection resulting in liver localization or the requirement for local delivery (i.e., intramuscular injection).

In this work, we have developed a quantitative, label-free mass spectrometry-based proteomics workflow to probe the nano-bio interface of soft lipid nanoparticles. We provide clarity on the experimental challenges of separating protein-nanoparticle complexes from free proteins present in bulk biofluids. Using this method, we investigate how altering the (i) ionizable lipid within the lipid nanoparticle and (ii) mRNA cargo impact the formation of the protein corona in (iii) various relevant delivery bioenvironments. These techniques are readily translatable to study other soft nanoparticle-based systems. By elucidating these fundamental protein-nanoparticle interactions, we can rationally tune the design of these mRNA-based biotechnologies for improved translation to clinical practice.

331c Desolvent-Free Vaccine Platform to Enhance Effectiveness of Protein Subunit Vaccines

Jaeyoung Park, Julie Champion

Vaccination has provided the most effective protection against viruses. Subunit vaccines have significant advantages over whole killed or inactivated viruses in terms of immune specificity and safety. However, protein subunit vaccines can have challenges such as low immunogenicity. Previously, our lab overcame this challenge by development of protein nanoparticles, developed by desolvating whole protein antigens and stabilizing via covalent crosslinking. Protein nanoparticles have been made from several antigens, including the highly conserved domains of influenza proteins such as the stalk of hemagglutinin and nucleoprotein. While the particles induce protective humoral and cellular responses, the desolvent, typically an alcohol, compromised the structure of the antigen. It is particularly important to preserve conformational epitopes in protein nanoparticle antigens in order to engage B cell receptors directly and enhance the strength and breadth of the humoral response. We hypothesized that effectiveness of vaccine protein nanoparticles could be improved by developing desolvent free fabrication processes that maintain antigen structure. Desolvent-free nanoparticles were synthesized by using ovalbumin as a model antigen via different approaches. Directly crosslinking or salt precipitating ovalbumin were identified as the best methods in terms of physicochemical properties (size~ 250 nm and PDI~ 0.2-0.29). These methods were also confirmed to produce hemagglutinin particles, an antigen that was unfolded by the desolvation process. Retention of ovalbumin conformation in desolvent-free nanoparticles was observed by circular dichroism while conformation of ovalbumin in desolvated nanoparticles was significantly compromised. The same phenomenon was also confirmed by molecular dynamics simulation of ovalbumin folding in the alcohol desolvent mixture compared to water. To evaluate the effect of desolvent-free nanoparticles on immune system activation, dendritic cells were stimulated with desolvated and desolvent-free nanoparticles. Surprisingly, expression levels of CD86, a maturation marker of dendritic cells, elicited by desolvent-free nanoparticles were higher than those by desolvated nanoparticles. For in vivo efficacy evaluation of the desolvent-free vaccines, mice have been injected intramuscularly with desolvent-free and desolvated nanoparticles, and also coated nanoparticles which were prepared by crosslinking ovalbumin proteins to the surface of desolvated nanoparticles. While coated desolvated nanoparticles were previously shown to better elicit immune responses than uncoated desolvated nanoparticles, the coated desolvated nanoparticles can only incorporate a limited amount of antigens with preserved structure and have not been directly compared to the desolvent-free nanoparticles. The results from in vivo study in concert with in vitro study will be presented and will reveal whether the new desolvent-free platform can be a promising strategy to improve immune responses to subunit protein vaccines.

331e Targeting Dysfunctional Blood-Brain Barrier Improves Nanoparticle Delivery into the Brain

Aria W. Tarudji, Hunter A. Miller, Evan Curtis, Brandon McDonald, Badrul Alam Bony, Alex Vecchio, Punita Dhawan, Forrest Kievit

Brain delivery of therapeutics is significantly hindered by the blood-brain barrier (BBB) that prevents the passive accumulation of materials in the brain. Nanoparticle (NP) delivery systems have been shown to improve brain accumulation and retention when there is severe disruption of the BBB such as in traumatic brain injury or brain cancer. However, delivery is still limited in disease states where BBB disruption is much less severe such as concussion and progressive neurodegenerative diseases observed during aging. Recent evidence suggests the tight junction protein claudin-1, which is not expressed in normally functioning brain endothelium, is increased on the brain endothelium during disease. Therefore, we hypothesized that targeting claudin-1 would allow for increased brain delivery specifically within dysfunctional regions of the BBB. PEG coated ultrasmall (3.5 nm) gadolinium NPs were modified with the claudin-1-targeting peptide C1C2. These NPs had approximately 19 peptides per NP with a coating density of around 1 peptide per 2 nm^2 of NP surface area. We found increased NP accumulation in brains of 12-month-old mice (corresponding to middle age in humans) as compared to 2-month-old mice as determined through MRI and confocal imaging. This increased accumulation correlated with increased expression of claudin-1 on brain endothelial cells in the 12-month-old mice. Additionally, we observed increased accumulation of these claudin-1-targeted NPs following traumatic brain injury, which correlated with increased claudin-1 expression. Our results strongly suggest claudin-1 as a strong target for improving delivery of nanoparticle-based therapeutics into the brain when severe disruption of the BBB is not present.

331f Neonatal Pharmacokinetics and Biodistribution of Polymeric Nanoparticles

Nuo Xu, Megan Wong, Elizabeth Nance

Title: Neonatal pharmacokinetics and biodistribution of polymeric nanoparticles

Presenting Author: Nuo Xu, University of Washington, US

Co-Authors: Megan Wong, Elizabeth Nance

Introduction: Therapeutic development for pediatric use has advanced in the last few decades, yet the off-label use of adult medications in pediatrics remains a significant clinical problem. The development of therapeutics for pediatrics is challenged by the lack of pharmacokinetic (PK) data in the pediatric population, a gap in data even more significant for neonates. Additionally, the role of nanomedicine, which can improve PK profiles of many therapeutics, has prominently focused on applications in adults. The US and European based pediatric formulation initiatives to investigate nanomedicine formulations for pediatric use necessitates greater understanding and greater data availability of nanomedicines in the pediatric population. In this study, we quantify the PK profiles and biodistribution of polymeric nanoparticles in neonatal rats. We focus on poly(lactic-*co*-glycolic acid)-poly(ethylene glycol) (PLGA-PEG) nanoparticles, which play an important role in drug delivery and have been widely studied in adults. Our results provide a guide for nanotherapeutic use in neonates.

Methods: PLGA-PEG (45k:5k, LA:GA=50:50) nanoparticles were formulated using standard nanoprecipitation methods in either polysorbate 80 (P80) or Pluronic® F127 (F127) (1). The fluorescent dye CF647® was conjugated to the PLGA backbone prior to formulation. PLGA-PEG particles were characterized using dynamic light scattering, and then intraperitoneally injected into postnatal (P) day 10 Sprague Dawley rats, with n = 4 (2 males, 2 females) pups per time point. P10 rats are term equivalent to a human. Rats were sacrificed at 30 min, 1 h, 4 h, 8 h, 24 h and 72 h after administration. Organ-level and tissue-level biodistribution were characterized via UV-Vis spectrometry and confocal microscopy, respectively. PK profiles of PLGA-PEG nanoparticles in serum, brain, heart, lung, liver, spleen and kidney were calculated from a calibration curve for each tissue on UV-Vis. Tissue from pups not injected with particles served as controls.

Results: We have previously shown PLGA-PEG nanoparticles made in F127 (PLGA-PEG/F127) and made in P80 (PLGA-PEG/P80) have different fates within the developing brain (1). PLGA-PEG/F127 nanoparticles were 60.2 ± 0.8 nm, -2.6 ± 0.3 mV, and had a PDI of 0.18 ± 0.01. PLGA-PEG/P80 nanoparticles were 66.2 ± 1.4 nm, -1.98 ± 0.3 mV, and had a PDI of 0.19 ± 0.01. Following i.p. injection, serum concentration of PLGA-PEG nanoparticles was present at 1h and continued to increase and reach the maximum concentration by 4 h, with a half-life ($t_{1/2}$) of 5.89 h for PLGA-PEG/F127. The accumulation of PLGA-PEG nanoparticles in brain, heart and lung also reached the peak at around 4 h. Liver, spleen and kidney had a delayed particle accumulation in the first 4 h after injection and then reached peak concentrations at 24 h – no to minimal concentrations were detected at 72 h. Among all organs, liver accumulated the most nanoparticles, whereas heart, lung and kidney had relatively low accumulation (less than 5% injected dose), and spleen was intermediate. Additionally, PLGA-PEG nanoparticles formulated with different surfactants had different biodistributions, either at the tissue-level or organ-level. $t_{1/2}$ for PLGA-PEG/P80 was shorter than that of PLGA-PEG/F127. At the tissue level, PLGA-PEG/P80 accumulated in neurons and microglia in the brain, but PLGA-PEG/F127 remained associated with brain capillaries. There were minimal differences in distribution in other organs, suggesting a unique mechanism of PLGA-PEG/P80 interaction with the brain endothelium (2).

Conclusion: PLGA-PEG nanoparticles are very promising drug delivery vehicles for the neonatal population, especially in the field of brain disease and injury, where the dosage determination depends on the development of PK profiles of nanoparticles in neonates. Understanding the PK profiles and biodistribution, and associated residence time, of PLGA-PEG nanoparticles in neonates will help design therapeutic dosing regimens with maximum efficiency and minimum toxicity, and thus help improve the speed of clinical translation and drug safety of nanodrugs into this underserved population.

References: (1) Joseph A. et al, *Biomaterials*. 2021, 277: 121086. (2) Kreuter J. et al, *J Drug. Target.* 2002, 10(4): 317-325.

Presenter biography: Nuo Xu is a first-year PhD student in Chemical Engineering department of University of Washington in Seattle, US. She earned a MS at University of Washington. She carries out research in the fields of drug delivery, neonatal disease, and neuroscience.

Learning Objectives: (1) Understand the PK profile and biodistribution of PLGA-PEG nanoparticles in term-equivalent rats. (2) Evaluate the differences in PK profile and biodistribution of PLGA-PEG particles made in different surfactants.

331g Combined Radiation-Induced Photodynamic Therapy and Immunotherapy Using Calcium Tungstate Nanoparticles, 5-Aminolevulinic Acid, and Epacadostat

Dhushyanth Viswanath, Sandra Torregrosa-Allen, Haley Harper, Bennett D. Elzey, You-Yeon Won

Head and neck squamous cell carcinoma (HNSCC) is the 7th most common cancer worldwide with a 5-year survival rate of around 50%. It is predominantly diagnosed at the locally advanced stage, for which the chemotherapy-radiotherapy combination ("chemoradiation", CT-RT) is the primary form of treatment. However, development of resistance to these treatments results in high rates of cancer recurrence and unfavorable prognosis. Therefore, there is need for alternative treatment modalities that use different mechanisms of action.

Photodynamic therapy (PDT) is one such option that has shown promise in achieving complete responses in patients with residual or recurrent HNSCC following conventional RT. The combination of RT and PDT (RT-PDT) has demonstrated improved outcomes for a multitude of cancers, with their synergism further bolstered when the two therapies are applied concurrently rather than sequentially. However, conventional concurrent RT-PDT is limited in three important ways; (1) an optical device is required to deliver light to the site of PDT; (2) conventional PDT is limited to surface tumors due to the small penetration depth of visible light in human tissue; (3) it is highly localized and cannot treat metastasis.

This talk will discuss our recent work on a new concurrent RT-PDT method that addresses the limitations of current RT(-PDT) approaches for treatment of HNSCC. This method uses a new RT-to-PDT transducer, namely, radio-luminescent $CaWO_4$ (CWO) nanoparticles (NPs) encapsulated within micellar self-assemblies formed by poly(ethylene glycol)-poly(lactic acid) (PEG-PLA) block copolymers ("PEG-PLA/CWO NPs"). These NPs are used in combination with 5-aminolevulinic acid (ALA), an FDA-approved endogenous water-soluble prodrug that causes preferential accumulation of an endogenous photosensitizer, protoporphyrin IX (PPIX), in cancer cells. Under ionizing radiation, CWO NPs emit UV-A/blue light, which activates PPIX. A confocal microscopy imaging study suggests that the radio-luminescence energy transfer process potentiates a type-II photodynamic reaction which converts molecular oxygen into cytotoxic reactive oxygen species (ROS), particularly, singlet oxygen ($^1O_2$) which oxidizes organelle and cytoplasmic membranes. *In vitro* clonogenic cell survival assays (performed using multiple cell lines having varying radiation sensitivity characteristics) confirmed that concurrent CWO NPs indeed significantly enhance the cell kill effects of X-rays. MTT assay results show that PEG-PLA/CWO NPs do not produce any dark toxicity in cells.

Furthermore, we are exploring the possibility of combining RT-PDT with immune inhibition blockade immunotherapy (IT) using Epacadostat (IDOi). IDOi is an inhibitor for indoleamine 2,3-dioxygenase (IDO), an enzyme that promotes T cell anergy in the tumor microenvironment and is frequently overexpressed in HNSCC. The hypothesis is that IDO inhibition coupled with localized cell death due to RT-PDT promotes T cell growth and maturation within the tumor microenvironment, resulting in systemic anti-tumor immune responses. A study is ongoing to validate the efficacy of the RT-PDT-IT combination therapy *in vivo*.

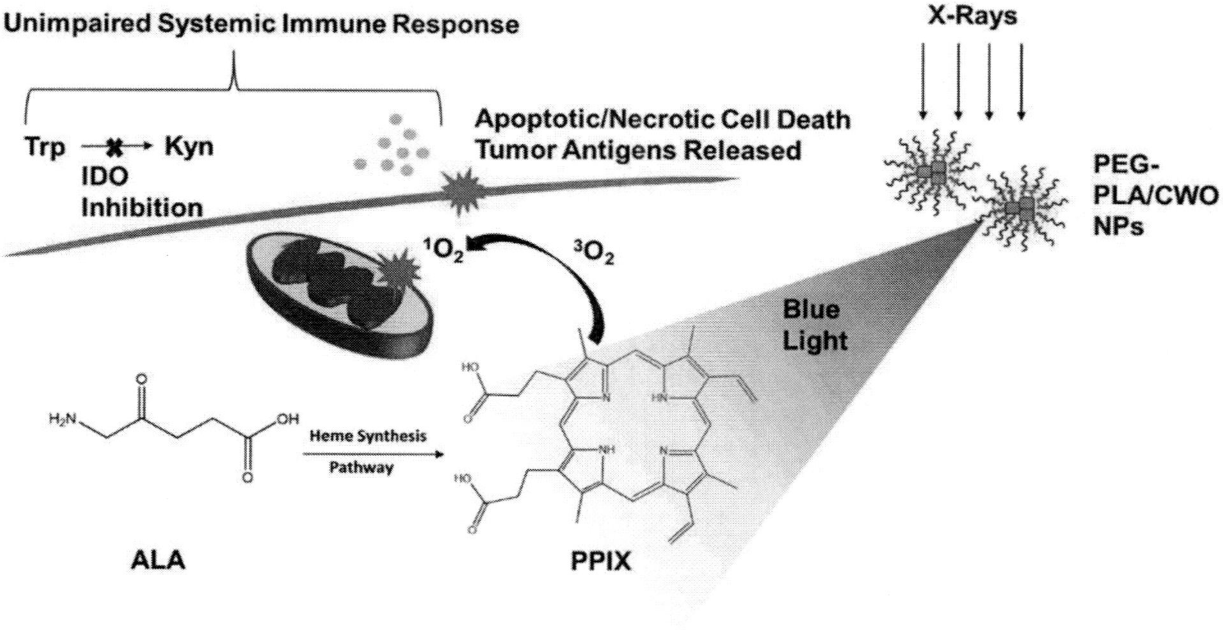
Unimpaired Systemic Immune Response
Trp Kyn
IDO
Inhibition
Apoptotic/Necrotic Cell Death
Tumor Antigens Released
X-Rays
PEG-
PLA/CWO
NPs
$^1O_2$
$^3O_2$
Blue
Light
Heme Synthesis
Pathway
ALA
PPIX

331h Modified Peroxamide-Based Reactive Oxygen Species (ROS)-Responsive Doxorubicin Prodrug

Joo-Youp Lee, Mina Jafari, Vishnu Sriram

Toxic side effects are common with systemic chemotherapies and impair the successful completion of treatment for many patients with cancer. Doxorubicin (Dox) is commonly used to treat various cancers including breast, bladder, lymphoma, and leukemia, but ~4-36% of patients experience potentially lethal cardiotoxicity in a dose-dependent manner. Although nanoparticle (NP)-based therapeutics have made tremendous progress, it is still a great challenge to meet the goals due to the two desired but conflicting concepts of sustained and controlled release. A stimulus-responsive linker approach can address these challenges by deactivating potent but toxic drugs upon conjugation to a polymer and reactivating them upon release triggered by a stimulus-responsive mechanism. Reactive oxygen species (ROS) can be used as a powerful internal stimulus, as the ROS level in cancerous cells (e.g., ~100 mM) can go up to ~5,000 times higher than those in healthy cells (e.g., ~20 nM).

Here, we present two ROS-responsive Dox prodrug-based NP systems that can deliver Dox in a sustained manner during systemic circulation while achieving controlled drug release at tumor sites. Our ROS-responsive Dox prodrug was physically encapsulated in polylactic acid (PLA)-polyethylene glycol (PEG) (PLA-PEG) NPs or chemically conjugated to PEG-poly-L-lysine (PLL) (PEG-PLL). The NP designs showed average sizes of <~100 nm with very stable Dox release profile in PBS at pH 7.4. While both NP designs showed ROS-responsive Dox releases up to ~70-80% within 48 hrs under 50-100 mM H_2O_2 and horseradish peroxidase (HRP) concentrations, the physically-encapsulated NPs showed faster Dox release in the first few hrs. The surface modification of NPs with hyaluronic acid (HA), a natural polysaccharide that can specifically bind to CD44, resulted in an increase in cellular uptake and efficacy against MDA-MB-231 breast cancer cell line. The therapeutic efficacy and off-target toxicity of our NPs were also investigated *in vivo* using an orthotopic xenograft tumor model using the MDA-MB-231 cell line.

AUTHOR INDEX